Drifting on Alien Winds

Exploring the Skies and Weather of Other Worlds

Drifting on Alien Winds

Exploring the Skies and Weather of Other Worlds

Written and illustrated by Michael Carroll

Michael Carroll
6280 W. Chesnut Ave.
Littleton, CO 80128
USA
cosmicart@stock-space-images.com

ISBN 978-1-4419-6916-3   e-ISBN 978-1-4419-6917-0
DOI 10.1007/978-1-4419-6917-0
Springer New York Dordrecht Heidelberg London

Library of Congress Control Number: 2011921341

Printed on acid-free paper

Springer is part of Springer Science+Business Media (www.springer.com)

Since 'tis certain that Earth and Jupiter have their Water and Clouds, there is no reason why the other Planets should be without them. I cannot say that they are exactly of the same nature with our Water; but that they should be liquid their use requires, as their beauty does that they be clear. This Water of ours, in Jupiter or Saturn, would be frozen up instantly by reason of the vast distance of the Sun. Every Planet therefore must have its own Waters of such a temper not liable to Frost.

Christian Huygens (1629–1695)

Preface

SOMETHING IN THE AIR

"Have you heard about that new restaurant on the Moon? The food's pretty good, but the place has no atmosphere."

It's an old joke, but it points out how important atmosphere is to us. It's no wonder – we live on the seafloor of a great ocean of air. This ocean is roughly 90 miles deep, with currents and eddies, whirlpools and waves. At our depth, the ocean above and around us presses down on each square inch of our bodies with a weight of 14.7 lb.[1] It's a brew of nitrogen, oxygen, carbon dioxide, some rarer gases, a bit of pollen and smog thrown in for good measure, and a substantial amount of water vapor. The humidity is responsible for hail, sleet, rainfall, and snowstorms, all of which end up as a different kind of ocean defining the shores of our continents.

We are used to this ocean around us, used to its weather patterns, storms, and seasons. Earth's atmosphere is unique. No other world in our Solar System has anything like it. But there are other oceans of air surrounding distant worlds, and these oceans would seem truly alien to us. In fact, the weather on other worlds is terrifying, inspiring, and baffling. Lightning bolts sizzle through Jupiter's atmosphere, powerful enough to run a small town for days. Hydrocarbon showers fall on Saturn's moon Titan, and sulfuric acid rains down on Venus. Snows drift from carbon dioxide clouds on Mars and methane ice crystal hazes on Neptune, where blue storms the size of Earth come and go in a matter of months.

Will we ever see these places up close? Will we fly on their winds or float among their thunderheads? Engineers are drawing up plans to do just that. In this book, we will journey with them. We'll travel, vicariously, to the planets and moons blanketed by substantial air, ones that generate dramatic weather. These include Venus, Earth, Mars, Jupiter, Saturn, Uranus, Neptune, and Saturn's planet-sized moon Titan. We'll take a side trip to Neptune's moon Triton, a world with a weather system somewhere between Earth's and the hard vacuum of space. Along the way, we'll meet inventors and their inventions designed to tell us about planetary atmospheres. We'll chat with scientists on the cutting edge of today's research, and take inventory of past and present explorations. Finally, we'll look at future possibilities: What is to

---

1. This is the pressure at sea level. Pressure in Denver, CO – the "Mile-High City" – is roughly 12.2 psi, and on top of Earth's highest mountains it hovers at 4.4 psi.

come, and how will it benefit us? What advanced probe designs are on the table, and what's already on the factory floor?

So sit back, relax, and enjoy a ride through alien skies.

Littleton, Colorado, USA   Michael Carroll

About the Author

Science journalist, writer, and artist Michael Carroll has been looking at the clouds for half a century. His twenty-five years as a science writer have afforded him the opportunity to work with many in the planetary science community, with contacts spanning from government research facilities to universities to aerospace corporations. Aerospace runs in his family; his father was an aerodynamic engineer for Martin Marietta, and his grandfather was both a general in the U.S. Air Force and a personal friend of Orville Wright.

Carroll is a Fellow of the International Association for the Astronomical Arts, and has written articles and books on topics ranging from space to archaeology. His articles have appeared in *Popular Science, Astronomy, Sky & Telescope, Astronomy Now (UK)*, and a host of children's magazines. His earlier book for Springer is *The Seventh Landing*, an exploration of our plans to return to the Moon. Carroll's twenty-some titles also include *Alien Volcanoes* (Johns Hopkins University Press), *Space Art* (Watson Guptill/Random House), and the children's book *I Love God's Green Earth* (Tyndale).

Carroll has done commissioned artwork for NASA and the Jet Propulsion Laboratory. His art has appeared in several hundred magazines throughout the world, including *National Geographic, Time, Scientific American, Smithsonian, Astronomy, Sky & Telescope, Ciel et Espace*, and others. One of his paintings is on the surface of Mars – in digital form – on the deck of the Phoenix Lander, and another was flown aboard Russia's MIR space station. Carroll is the 2006 recipient of the Lucien Rudaux Award for lifetime achievement in the Astronomical Arts. He lives with his artist/sometimes-coauthor wife, Caroline, in Littleton, Colorado.

Acknowledgements

My thanks, first, to Maury Solomon and the good people at Springer, for encouragement and guidance in making this book all that it could be. Thanks also to Phillip J. Weisgerber and Aldo Spadoni for insights into Northrup Grumman's involvement in the history of planetary exploration. Pat Rawlings and Harvey Feingold of Science Applications Institute Corporation helped me track down early Titan blimp concepts. Marty Caniglio of Denver's 9NEWS provided creative aid in my understanding of weather on our own planet. (Forecasters in Colorado have to be on their toes!) Catherine Decesare at Colorado State University and Marc Levine of the Denver Museum of Nature and Science offered invaluable insights into Mesoamerican history and religious beliefs. Carolyn Porco spent long e-mail sessions with me so I could get my Saturn ring painting close to reality. Mark Wade provided the fine Mars 6 image and other Soviet space advice. Richard Kruse helped out with beautiful VEGA balloon images. Peter Burns and Dave Welch were instrumental in my obtaining historical art and permissions for early planetary probe designs from Martin Marietta. Ron Miller let me use his own image of the Venera aeroshell, which he took all by himself in Moscow, and helped me track down some classic space art. As always, Don Mitchell is "the bomb" when it comes to reprocessed Soviet/Russian imagery; thanks for graciously sharing your beautiful work, Don! My thanks to Ursula Schafer-Simbolon of the Historical Archive at Deutsches Technikmuseum, Berlin (SDTB), for the Lilienthal image. My thanks also goes to the editors of *Astronomy*, *Sky & Telescope*, *Scientific American*, and *Astronomy Now* magazines for letting me use earlier research – done for their fine publications – to further this book. A special thanks goes to my mother, Edith Carroll, for transcribing sometimes technical and mumbly interviews, to my daughter Alexandra, for moral support and transcription of other interviews, and to my wife Caroline, who tirelessly edited long stretches to make this a friendlier book. Finally, to my father, Pat Carroll, for access to his collection of space history, both printed and experiential.

Contents

Part I
Starting Here and Getting There

Chapter 1

The Sky at Home

Sunsets on two planets: (left) Orange sunset in Colorado, Western Hemisphere, Earth (photo from the collection of Michael Carroll) and (right) Blue sunset at Columbia Hills, Gusev Crater, Mars (MER image NASA/JPL)

The skies above us offer a dynamic theater of shifting colors, boiling thunderheads, scatterings of ragged white patches adrift on a changing cerulean canopy, beams of light, spectral halos, and flashing

M. Carroll, *Drifting on Alien Winds: Exploring the Skies and Weather of Other Worlds*, DOI 10.1007/978-1-4419-6917-0_1, © Springer Science+Business Media, LLC 2011

lightning. This is the sky of Earth. Although we will see hints of our weather on other worlds, our skies are unique. Before we can truly understand alien skies, we need to first understand Earthly weather and what makes it tick. In an odd cosmic twist, by studying and comparing our weather to others we'll actually learn about how our own world works.

MAKING WEATHER, EARTH STYLE

Ever since sparks drifted skyward from Cro-Magnon fires, people have understood that hot air rises. This simple fact of life is a vital component of what makes our skies behave in the way they do. Heat rises, cold air sinks, and weather is born.

If the physical universe has a goal, it is to keep things even. Nature abhors a vacuum; just ask any astronaut. When you take a spacecraft into orbit and throw the door open, the air inside will rush out. It's nature's attempt to keep the pressure inside the same as the pressure outside. Dry regions pull humidity from moist areas in an effort to evenly distribute the water vapor. Cool cream rapidly spreads through a hot coffee cup, transforming your cuppa from black to sienna. Here on Earth, our weather is simply a mechanism in constant search for equilibrium in our planet's temperatures. Achieving this is not easy; temperature is in constant flux because of the tilt of our world, which alternately points one hemisphere – and then the other – toward the Sun at any given time.

Nature makes this point well by providing us with a comparison next door. In the cosmic scheme of things, our Moon orbits the Sun at the same distance as Earth does. But compared to our home planet, temperatures on the airless Moon swing wildly. When Apollo astronauts stood on the lunar surface facing the Sun, the front of their suits simmered at water's boiling point, while their backsides dropped to a frigid −279°F. In a very real sense, their pressure suits served as miniature environments designed to even out those temperatures. Just as Earth's atmosphere distributes temperatures fairly evenly, radiators inside spacesuits pump liquid throughout the suit, cooling the hot spots and keeping the shadowed areas from freezing solid.

On the day side of our world, hot air forms over the region of the planet nearest the Sun. On Earth, this area is near the equator. Here, hot air rises, moves away from the equator, and drifts to the poles. It cools at high altitude and sinks

Astronaut Mike Fossum works at the International Space Station. Like Apollo suits before them, shuttle/ISS space suits provide miniature environments to even out temperatures for astronauts, just as the atmosphere evens out temperatures across Earth (STS124/International Space Station/NASA)

S124E006335

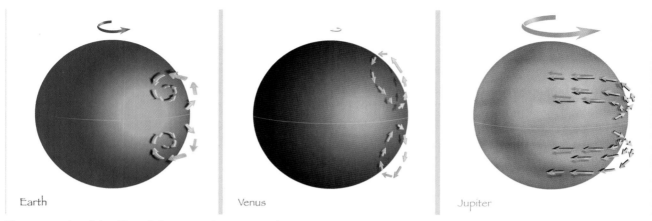

Three examples of the effect of planetary spin on air circulation

back down, migrating at low altitude back toward the equator. This circulating carousel of atmosphere is called a cell. Cells are arranged like donuts encircling the planet parallel to the equator. A simple model of these cells can be found on the planet next door, Venus. There, air rises above the equator, flows toward the poles, where it cools, then returns again to the equator, completing an airy conveyor-belt from the planet's waist to scalp. Venus turns very slowly, so its motion does not disturb the air currents.

Earth is nearly identical in size to Venus. The diameters of the two worlds differ by only 406 miles; Earth is the slightly larger of the two.[1] If Earth were a stationary or leisurely-turning sphere, the Hadley cells could go about their business merrily distributing warm air from equator toward the poles and cool air back toward the equator, as the air does on slow-moving Venus. But life is not so simple here. Earth completes a full rotation, turning toward the east, once each 24 h.[2] This motion drags the air above it westward, twisting the clouds into the familiar spirals of tropical storms and hurricanes, a phenomenon called the Coriolis effect. Earth has six zones of air circulation, three in the northern hemisphere and three in the south. The largest zones, called Hadley cells, extend from the equator to 30° north and south. Above these, twin mid-latitude cells, known as the Ferrel cells, extend to 60° north and south, and encircling Earth's north and south polar regions lie the polar cells.

As air from the Hadley cells rebounds from the north, its motion pushes westward and becomes the trade winds. British astronomer Sir Edmund Halley[3] was the first to try to explain the trade winds. He proposed, correctly, that solar heating forces air over the equator to rise, and that this air is replaced by cooler air from north and south. But the east-west motion of the

The circulation of the atmosphere on Venus is a simplified version of what we find on Earth

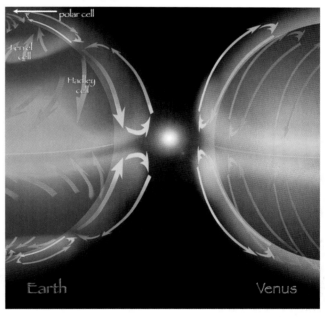

1. Venus measures 7,520 miles, or 12,103 km. Earth's diameter spans 7,926 miles, or 12,756 km.

2. Seen from above the North Pole, this motion is counterclockwise, or prograde. Some planets spin the opposite way, in what is called retrograde motion.

3. The same Halley whose name is associated with the most famous periodic comet, Halley's Comet.

winds stumped Halley. After all, if the atmosphere was being driven away from the equator by solar heating, the winds should blow north and south, but the trades had a decidedly westerly component. Why? He theorized that the Sun heated the air inconsistently over the course of each day, setting up temperature changes that drove the winds. This was a fine hypothesis, but reality turned out to be more complex.

The solution was to come later, but the mystery was an important one. The workings of the wind posed more than an academic question. For mariners, it could be a matter of life and death. Although the trade winds beneath the Hadley cells blow reliably, air along the equator can be stagnant. Hot air rises here in two walls, moving toward the poles along the Hadley cells. The belt of sea air girding the equator between the rising air has high humidity, low pressure, and often little or no wind to fill sails. This belt is called the Doldrums, and seafarers in bygone days spoke of it with dread. It is a miserable place for a sailing ship to be stranded. Coleridge described it as being "as idle as a painted ship on a painted ocean."[4] Between 30 and 35° north of the equator, and in a similar region in the southern hemisphere, lies another dismal region for mariners where many have perished. Called the Horse Latitudes, it's a swath of ocean where the breezes stall, leaving sailing ships becalmed for days or even weeks. The static air results from a high pressure area between the Hadley cell along the equator and the Ferrel cell to the north. Ancient mariners, who relied on winds for safe passage, often ran low on supplies during the calm times. Legend has it that the name originally came from Spanish ships stranded while transporting horses to the West Indies. Stuck with no wind power for weeks on end, it is said that crews would throw their horses overboard in an effort to save water rations.

In 1735, British barrister and meteorologist George Hadley published a new theory that took into account Earth's rotation as the element that was forcing trades in their westerly direction. The weather mechanism bears his name to this day.[5] Hadley's theory didn't solve all the westerlies' mysteries, however. The trade winds seemed to be twice as powerful as Hadley's theory projected, and they moved in different directions. Hadley's model was too simple. It involved only one cell over the tropics. He didn't know about the Ferrel and polar cells to the north of the Hadley cells. The global wind patterns predicted by Hadley's cells do not, in fact, exist. It took over a century – and some fancy mathematics – for a complete model to emerge that could explain the nuances of Earth's winds, breezes, and eddies.

The winds of the world take on personalities of their own. Egyptians and others from the Middle East to the Arabian Peninsula suffer the *Simoom*, Arabic for "poison wind" The seasonal gale sucks moisture from the air and causes the temperature to spike. Europe's Mistral and Marin winds have more positive associations with welcome rains and warmth in cold weather. Monsoons – winds that reverse with the seasons – regularly bring torrential rains to Asia, Australia, and western sub-Saharan Africa. In the late summer, a partial or incomplete monsoon visits North American desert areas in Mexico and the southern United States, but this monsoon is not cyclical, like

4. From *The Rime of the Ancient Mariner* by Samuel Taylor Coleridge.

5. Apollo 15 landed in a mountainous area on the Moon named after Hadley. A crater in the rugged highlands of Mars also bears his name. Fittingly, the crater is just south of the equator.

the ones in the eastern hemisphere. Earth's winds become part of life as they come and go each year, ushering in some seasons and bidding others farewell.

Wind is an important part of Earth's weather systems, not only for its role in distributing heat but also in its role as sculptor. The mountains, plains, and valleys of our planet are carved by wind and the rain that winds bring. Aeolian (wind) erosion is dominant in dry regions, where windblown dust and sand wear down rock surfaces and carve hollows in canyon walls. The wind ultimately drives precipitation, and that precipitation is the prime mover of material across any earthly landscape. In fact, says planetary scientist Jani Raderbaugh of Brigham Young University, "On Earth, the dominant land forms are river channels. You get mountain peaks crammed together, rivers carving through and transporting material to the plains." So, on Earth, the world of weather above helps shape the land of the stony world below. We will also see this dramatic effect on Mars and Saturn's moon Titan.

Mistral wind blowing near Marseille, France. In the center is the Chateau d'If (photo by Vincent, Wikimedia Commons)

Winds of the World

Many of the world's winds appear at predictable times and in predictable ways. Here are just a few.

Bora: The fickle Bora gusts across Croatia and the east coast of the Adriatic Sea. The Bora is a cold winter wind. High-pressure air backs up behind the snow-covered Dinaric Alps. Warm air from the Adriatic Sea mixes with the chilled mountain air, rushing across Montenegro, Bosnia/Herzegovina, Croatia, and Serbia. The "light Bora" (Bora chiara) blows through clear skies. The "dark Bora" (Bora scura) carries rainstorms.

Chinook: These winds blow across western North America, where the Canadian Prairies and Great Plains meet the Rocky Mountains. Pacific Northwest Native Americans originally coined the term to mean a warm, moist, coastal wind, but Chinook has come to mean a warming wind. A strong Chinook can melt snowdrifts and raise temperatures by 50° in a matter of hours.

Foehn wind (or *föhn*): This is a dry downslope wind in the lees of mountain ranges in Central Europe. Warming air rising on the windward side of the mountains drops its moisture. Dry winds on the leeward slopes warm, raising temperatures by as much as 30°C (54°F) in just a matter of hours, giving the wind its nickname of "snow-eater."

Khamseen or *Khamsin*: This is a strong, hot wind that carries dust through North Africa and the Arabian Peninsula. Khamsins are triggered by low-pressure areas moving east along the southern Mediterranean and North African coasts from February to June. In Egypt, Khamsin arrives in the spring, carrying sand and dust from the deserts.

Marin: This blows from the southeast across France's Gulf of Lion, bringing warm moist air to the coasts of Languedoc and Roussillon. It also brings rain and fog to the Corbieres, Montagne Noire, and the Cevenne regions. When the wind is gentle, it can usher in a wave of tourists swimming in the gulf of Lion by Marseille,

(continued)

Winds of the World   (continued)

but the Marin is capricious, sometimes triggering heavy, dangerous surf along the coast.

Mistral: The mistral is an example of a katabatic wind, a wind generated by the difference in pressure between warm and cold regions. High pressure in the Atlantic or northwest Europe interacts with low pressure in the Mediterranean. The Mistral rushes south through the Rhone Valley. The Alps and the Massif Central funnel the breezes into strong winds. In France, the mistral is most dramatic in Provence, Languedoc, and down the Rhone Valley from Lyon to Marseille, stretching as far as Corsica.

Monsoon: This is a cyclical wind that results in strong seasonal rains. Classic monsoons descend upon Asia, northern Africa and Australia at different times of the year, depending on their location. Partial monsoons occur in southwestern regions of North America.

Simoom: This means "poison wind," and with good reason. This strong, dry, scorching wind appears abruptly, spreading across the Sahara, Palestine, Israel, Jordan, Syria, and the deserts of Arabian Peninsula. Its temperature may exceed 129°F, bringing great clouds of dust and sand. Because of its sudden effect, the Simoom has been known to cause heat stroke in both people and livestock.

Sirocco: This is a fierce Mediterranean wind that comes from the Sahara, roaring into North Africa and Southern Europe. It can reach speeds of 65 mph. The Sirocco comes in the fall and spring, causing dusty, dry conditions along the northern coast of Africa, storms in the Mediterranean Sea, and cold, wet weather in Europe.

A NEW SPIN ON THINGS

Seasonal changes exert a critical influence on our weather. A common misconception is that seasons are linked to Earth's distance from the Sun, but Earth's orbit is nearly circular. Seasons are actually driven by the tilt of the axis around which the planet spins.

The axis of Earth tilts 23.5° from the plane of its orbit around the Sun. At times, the axis tips the northern hemisphere toward the Sun during the day. The Sun's path arcs high up in the sky, with the Sun nearly overhead at noon. Half a year later, the northern hemisphere tilts away from the Sun during the day, so that the Sun appears to travel lower in the sky. (In the southern hemisphere, the seasons are reversed.) The subtle tilt makes for dramatic swings in temperature, partly because this slant also causes short winter days and long summer ones. The distance between Earth and the Sun has a very subdued effect on temperature compared to the seasonal effect, although it decreases temperature swings in the north. Long-term patterns of weather are known as climate, which is affected by the wobble of a planet's axis, composition of gases in the atmosphere, changing long-term patterns of the air's flow, and the shape of a planet's orbit (its varying distance from the Sun; for more on climate, see Chap. 6).

Mars shares a similar set of seasons to Earth's. At 25°, Mars has a nearly identical axial tilt to ours. Its day – the time it takes to spin around its axis once – is only a few minutes longer than Earth's. Early observers noted this as they viewed the Red Planet by telescope. Mars's earthlike tilt, and presumably earthlike seasons, reinforced the idea that Mars was similar to Earth in its climate and weather. Astronomers such as Schiaparelli, Antoniadi, and Percival Lowell drew maps displaying webs of dark lines across the face of Mars. The dark areas appeared slightly greenish, and as seasons changed, a "wave of darkening" moved across the dark zones, as if melting polar caps were watering vast jungles or mossy plains. Why not? The planet seemed tangled in a skein of straight lines that also darkened with the seasons. Leading scientists proposed that these lines might be channels carrying water, and Lowell led the camp arguing for an intelligence behind those channels, which they called "canals."[6] But despite an Earthlike set of seasons and length of day, other forces conspire to give Mars a quite alien environment (see Chap. 6).

Although Venus is Earth's twin in size, its seasonal changes are virtually non-existent. Venus's axial tilt is less than three degrees from vertical, compared with the plane of its orbit (a vertical axis would be either 0 or 180° of tilt). But since it spins in a retrograde, or clockwise, motion viewed from the north, scientists say that its axis is tilted all the way around to 177°.[7]

"'SCUSE ME, WHILE I KISS THE SKY…"

Earth's atmosphere has a structure unlike the atmosphere of any other world. Stacked like the layers of a wedding cake, Earth's canopy of air arranges itself in distinct tiers, each with gas molecules constantly drifting, mixing, and bouncing into each other. If we hitched a ride on a water molecule, we just might get to tour the entire cake.

The bottom-most layer of this gaseous pastry, the interface between solid surface and firmament, is called the troposphere. Our water molecule – a triad of two hydrogen atoms and one oxygen atom – floats easily through this dense region, sometimes joining others to become clouds, sometimes sticking together with many to fall as rain or snow. Our molecule slips into an updraft of warm air. As we ascend, clouds and sheets of rain parade by, interrupted now and then by clear air. In the distance, lightning flickers beneath a violent thunderstorm. We pass a commercial airliner making its way through the friendly skies. Reaching 6 miles up, the troposphere gets cooler with altitude, dropping about 4°F every 1,000 ft. A constant rise and fall of currents mixes the atmosphere well in this lowest of layers. The currents are generated from solar heating and from heat welling up from the ground. Cold air pours from above in a constant aerial cotillion. In fact, the Greek word *tropos*, from which the word troposphere stems, means "turning."

The troposphere is the realm of all the phenomena we know as weather. Eight tenths of Earth's atmosphere settles here. All the action in the Hadley cells takes place within the troposphere. The troposphere is composed primarily of nitrogen and oxygen, with small amounts of carbon dioxide[8] and other gases, as well as the water vapor on which we now ride.

From a viewpoint out in space, it is here that the great swirls of white cloud stand out against the blue of the oceans, so remarkable in images from our weather satellites, shuttle views, and Apollo snapshots. Clouds come in a wild variety of forms, each with its own genesis and unique effect on the environment. Three main shapes distinguish cloud types: cirrus, stratus, and cumulus. Their Latin nomenclature was originally suggested in

6. Early Italian observers referred to the straight Martian lines as "canali," meaning "channels." Although no alien builders were implied, the word made it to the English-speaking world as "canals," something clearly artificial.

7. Most planets in the Solar System rotate in a prograde motion. Viewed from the north, this motion is counterclockwise. All planets also orbit the Sun in a prograde direction. If a planet or moon spins the opposite way, its motion is retrograde.

8. Although carbon dioxide is a greenhouse gas, water vapor plays a far more important role in the trapping of heat within Earth's atmosphere, accounting for some 70% of the greenhouse effect.

The layered structure of Earth's atmosphere is obvious from orbit (NASA)

1803 by British meteorologist Luke Howard.[9] Cumulus, meaning "heap," describes the puffy clouds at low altitude. Stratus describes cloud formations that are layered. Cirrus comes from the Latin for "wisp" or "hair." Stratus clouds stretch into thin layers fairly near the ground. Above them float the puffy cumulus clouds. Stratocumulus clouds are fine, puffy clouds arranged in long flat layers. Cumulonimbus clouds – thunderstorms – often grow into the top of the troposphere. Sometimes they even break into the lower stratosphere. Clouds with the prefix "alto" are at mid altitudes. Altostratus are layered, while alto-cumulus are higher altitude puffs. Above them, in thin clear air, spread sheets of cirrostratus and cirrocumulus clouds. At the top, at low pressures where the last water vapor can condense into visible clouds, drift the ghostly cirrus clouds.

Cloud Language

When talking about clouds, meteorologists the world over use a common language: Latin. Here is a sampling of Latin labels associated with cloud types.

Castellanus: Shaped like the turret of a castle

Fibratus: Fiber-like; strands of cloud

Lenticularis: Lens-shaped

Mammatus: Breast-like, hanging pouch structure

Nimbus: Rain-bearing clouds

Undulates: Clouds arranged in waves, undulating

As our water molecule reaches the crown of the troposphere, we pass the last of the cirrus clouds. The air becomes calm. We have arrived at the tropopause, a stable ceiling averaging 6 miles high. The atmosphere actually breathes and undulates like a living thing, varying in height and density. The troposphere crests at about 4 miles high over the poles, but can expand up to 12 miles in altitude over the equator. At its top, the tropopause caps most clouds, forming an upper boundary to our weather. But the occasional violent thunderstorm breaks through, boiling up into the stratosphere above us. Our water molecule is too heavy to go any higher, but as often happens, solar radiation breaks us apart in a process called dissociation. Now, our two hydrogen atoms go their separate ways. We'll stay with the oxygen, just because it's nice to breathe.

Now well into the stratosphere, we float too high to be influenced by heat from Earth's surface. In an odd reversal, temperatures actually increase with altitude here. The stratosphere tops out at 30–40 miles. Our surrounding air has warmed from an average of −71°F at the tropopause to 5°F at the top. The temperature gradient tends to result in calm air, so commercial

9. Luke Howard, 1772–1865.

Altocumulus clouds drift
15,000 ft above the island of
Oahu (NASA)

Cumulonimbus clouds boil into the skies
over the Gulf of Mexico. Note how the cloud
structure flattens out as it reaches the
stratosphere (NASA)

Cirrocumulus clouds in the
Sahara Desert above Mali
(NASA)

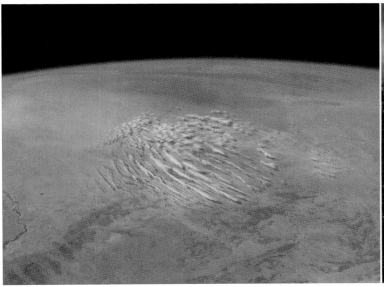

flights like to cruise in the lower stratosphere. The world's famous ozone layer, so critical to the health of our environment, resides in its upper regions. In fact, the ozone layer is responsible for the high-altitude heating. Ozone absorbs heat from the Sun's UV light to heat the stratosphere. Ozone also plays a role on other bodies we will visit. Jupiter, Saturn, Uranus, Neptune, and Titan all have stratospheres (with temperatures increasing with altitude), but the source of the heating from above is not yet understood. At these distant worlds, heat from sunlight is not strong enough to result in the temperatures – or structure – seen in their stratospheres.

For the next 20 miles, our ascent takes us through the mesosphere, where temperatures again fall with altitude. Most meteors[10] burn up in this region. Atmospheric friction causes the tiny particles of rock or dust to vaporize as they enter the atmosphere. A grain-sized piece of stone leaves an incandescent trail behind it spanning a few feet across, but many miles long. Only the rare larger meteoroids (larger than a marble) get through to the surface.[11] The mesosphere is the coldest layer of the atmosphere.

Ironically, we find the hottest layer of our atmosphere near the top. Above the mesosphere, the thermosphere stretches from 50 to 320 miles. As we continue our climb, temperatures begin a gradual increase once again, reaching 3,000°F at the summit. Above 320 miles, we watch auroras play across the ionosphere, the inner edge of Earth's magnetosphere where sunlight ionizes air molecules. We drift up into the exosphere, where we wander for long periods without bumping into another atom. The air is so rarified here that its amorphous boundary is considered the vacuum of space. Only chance will determine whether we bump into an atom that sends us groundward, or whether we continue our journey out into space.

10. The term "meteor" actually refers to the glowing trail in the wake of an incoming object. A meteoroid is a stony or metallic object that enters Earth's atmosphere. If it reaches the ground, it becomes a meteorite. If the object is larger than about 330 ft (100 m) across, it is considered an asteroid.

11. The exception to this size rule is a constant drizzle of microscopic meteoritic dust that drifts in from space. Its descent is so slow that it settles to the surface with little or no change from atmospheric effects.

The Oxygen Enigma

Any extragalactic traveler would immediately notice something unique about Earth. Our world is the only planet in the Solar System where you will find free oxygen as a substantial portion of the atmosphere. Oxygen is reactive; that is, it tends to combine with nearby gases or minerals to make other things. So, for example, we get water when oxygen hitches a molecular ride with hydrogen, or oxidation as oxygen combines with minerals in rocks.

Only one thing enables Earth to retain free-floating oxygen in such quantities: life. Living things transformed our environment 2.5 billion years ago, during what geologists call the Great Oxidation Event. In Earth's early history, the atmosphere consisted mostly of water vapor and carbon dioxide and may have been quite similar to Mars or Venus. But as life began to propagate in the seas, its exhalations contributed "pollution" in the form of pure oxygen. Before 3.5 billion years ago, the majority of the Earthly population were the blue-green algae and other single-celled microbes. These organisms took in carbon dioxide and used solar energy to break it down into carbon and oxygen. As carbon-based life forms, they incorporated the carbon into their tissues (in other words, they digested it) and tossed the oxygen away. Over a 1,000 million years, the little creatures changed our planet's atmosphere in a most dramatic way, drastically lowering the carbon dioxide levels and infusing the air with oxygen.

How do we know all this? The first hints came in 1969, when Canadian geologist S.M. Roscoe pointed out that some layers of ancient material (2.45 billion years old) that are easily oxidized were left unchanged.

Other sediments that tend to oxidize are rare or absent in the fossil record before 2 billion years ago. Other tests using sulfur isotopes confirm the fact that Earth's atmosphere had virtually no oxygen prior to 2 billion years in the past. But after that, the planet bloomed with innumerable oxygen-producing plants; Earth would never be the same. The Great Oxidation Event shifted our world from a carbon dioxide to a nitrogen/oxygen environment. It was a revolution, the most profound event in the history of our planet's atmosphere so far.[12]

Even today, most of our planet's oxygen comes from algae in the oceans. For this reason, exobiologists, or scientists who search for life in the cosmos, hope to use the presence of oxygen on distant worlds as an indicator that biology is active within those far-off environments.

12. It remains to be seen whether humankind will usher in the next great revolution, changing our atmosphere from oxygen-rich back to a mostly carbon dioxide environment. Our factories and cars seem intent on such a transition.

WHEELS WITHIN WHEELS

Humankind is becoming environmentally savvy. We stamp recycle symbols on our plastic and paper, and post bins outside convenience stores for our glass bottles. But Earth has been recycling for billions of years. Our planet weaves a complex tapestry of interlaced systems that contribute to the health and stability of our environment. Oxygen and carbon continually swap places. Water becomes vapor and rains back into our lakes and oceans. Even the rocks recycle constituents of the atmosphere.

Three of the planet's many cycles impart the greatest impact to our weather and living systems. The first is the most obvious: the water cycle. Ancient peoples were intimately familiar with it, as this cycle determined the success of crops. As the Hebrew writer Isaiah put it, "…the rain and the snow come down from heaven, and do not return to it without watering the earth and making it bud and flourish…"[13] The Mayan codex of Dresden dedicates a great deal of text to the rain god Chaac, and to monsoon seasons and floods. As early as 600 BC, the Greeks developed the first scientific theories of the comings and goings of water, which they considered the primordial construction material of the universe.

The water cycle is more complex than a matter of rain turning to vapor and then condensing into rain again. The world stores its moisture in many forms. Water floats in condensing vapor as clouds. It rests atop mountains as ice, snow, and glaciers. It sloshes through rivers, lakes, and oceans in liquid form. It even settles into subsurface lakes called aquifers and permanently frozen ice layers – called permafrost – under tundra regions. Solar heating works on all these sites, moving currents with its heat and urging liquids and ices to become vapor again.

Liquid water evaporates into vapor from seas, rivers, and lakes. Water also reaches the atmosphere through transpiration, as grasslands and forests breathe moisture into the air along with oxygen. As the water comes back down, it settles again upon snow-capped peaks, in rivers and oceans, and beneath the surface of the ground as it trickles down into aquifers. Bodies such as lakes and oceans are recharged by runoff from melting snows, by

13. Isaiah 55:10, New International Version of the Bible; written in about 760 BC.

surface runoff from precipitation, and by water from springs that vent liquid from underground sources. Then the cycle begins again, keeping waters clean and infused with oxygen from the air.

The second important cycle involves the interplay between oxygen and carbon dioxide. Every two millennia, plants take in all of Earth's atmospheric carbon dioxide, converting it into oxygen. It takes a thousand times as long (2 million years) for all the water on Earth to recycle from surface to air and back again.

Oxygen also cycles between the air and the solid surface. Metals within the rocks oxidize, combining oxygen with their minerals. Marine organisms grow shells of calcium carbonate, which contains oxygen. These shells eventually turn into limestone. The rocks beneath our feet hold one hundred times the amount of oxygen that the air does! Eventually, plants take in the minerals in the rocks, and free the oxygen through photosynthesis, and the oxygen cycle continues. Both oxygen and carbon dioxide replenish the atmosphere through another avenue: volcanoes. Earth's crust is fractured into large segments called plates. These plates float on a rubbery layer of semisolid rock, moving at about the speed that a human fingernail grows. Some plates ram into each other, pushing up mountain chains and causing earthquakes. Others subduct, or slide under each other. Here, the gases that were absorbed from the atmosphere and trapped in the rock are freed as the rock melts. Those gases, along with water vapor and molten rock, escape back into the atmosphere through volcanic eruptions.

A third cycle interfaces with the oxygen cycle. The carbon cycle mirrors the oxygen cycle but increasingly involves human impact on the environment. Earth's air-breathing creatures take in oxygen and breathe out carbon dioxide. Carbon dioxide is known as a greenhouse gas; it acts to trap solar energy as heat, warming the atmosphere. Water vapor has the same effect. Without these greenhouse gases, Earth would be considerably colder and lifeless. (For more, see Chap. 5.)

Holes in Your Ozone

Oxygen participates in another cycle at the top of the stratosphere. Near the top of the stratosphere, at about a 40-mile altitude, the air is thin, and solar radiation gets through more readily than in the thicker, lower layers. As we saw with our water molecule, solar ultraviolet radiation can tear things apart. Molecules of oxygen gas, which consist of two oxygen atoms, are no exception. When radiation breaks them apart, what's left are solitary oxygen atoms. Oxygen atoms don't seem to enjoy being alone; they are very reactive, combining readily with intact oxygen molecules to form a triple atom. This oxygen trinity is known as ozone. A layer of ozone drifts atop the stratosphere and acts as a planetary sunblock. Ozone forms an excellent shield, preventing solar radiation from getting to the ground where living things can be harmed by it.

Some types of pollution destroy ozone. NASA's *Total Ozone Mapping Spectrometer* satellite has charted the growth of a permanent hole in Earth's ozone layer over Antarctica. The hole steadily increased over the lifetime of the spacecraft, from 1981 to 1999. Evidence now points to a second hole developing over the northern hemisphere. Industrialized nations are working to curb the specific pollutants that directly affect the protective, but delicate, ozone layer. Researchers estimate that it will take decades, or even centuries, to undo the damage that has been done so far. Oddly enough, our studies of Venus gave us some of our first clues about this environmental crisis (see Chap. 5).

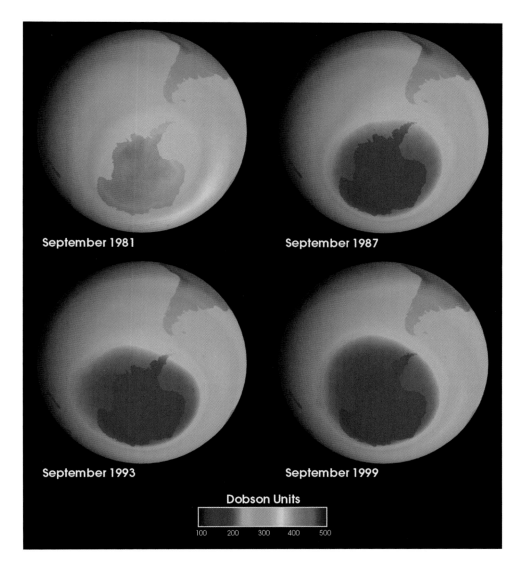

NASA's Total Ozone Mapping Spectrometer charted the growing ozone hole over the South Pole from 1981 to 1999 (image courtesy NASA/Visible Earth, http://visibleearth. nasa.gov/)

As plants die and decay, they release carbon dioxide back into the air. If plants or bacteria are buried deep enough, they become compressed and turn into oil, coal, and natural gas. These "fossil fuel" reserves are significant reservoirs of carbon dioxide. When they are burned, the carbon dioxide is released into the atmosphere once again. If all Earth's carbon dioxide was released into the atmosphere, pressures would rise to 90 times what they are now, and temperatures would soar. For a glimpse of what things would be like, we have only to look at the planet next door, Venus. With the onset of industrialization, humans seem intent on freeing the carbon dioxide stored in the ground. By studying other worlds, we are learning some valuable lessons about stewardship of our own environment, as we will see.

Aside from solar heating, there are other cosmic connections with Earth's cycles. Carbon-rich meteors, called carbonaceous chondrites, add carbon to Earth as they wander in from space. Meteorite types fall into two general groups. The most ancient contain elements similar to those found in the

Cycles on Earth are tied into nitrogen, carbon, and water. Percentages of these atmospheric constituents are quite different on our two nearest planetary neighbors

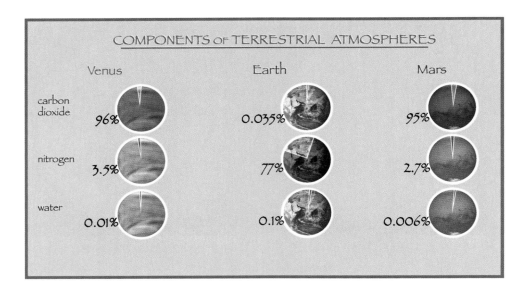

A Brief History of Earth

By today's standards, early Earth was a world as hostile and alien as Venus or Jupiter. Beginning as a molten globe, its crust cooled within a few million years of its formation some 4.6 billion years ago. The earliest atmosphere, left over from the primordial solar nebula, was rich in hydrogen and high in temperature. An eternal darkness enshrouded the searing world as dust from asteroid and comet impacts choked the air. An eerie glow infused the clouds day and night from hot rocks and rivers of lava. The hot hydrogen atmosphere could not last. The light gas floated away or was blasted into space by huge impacts during the initial bombardment, ending 3.9 billion years ago. Gases from Earth's interior finally replaced the hydrogen atmosphere, as volcanoes belched water vapor and carbon dioxide into a new sky.

Water condensed from the atmosphere. The rains began to fall, and the skies cleared. One of the oldest rocks ever found on Earth is a 3 cm wide chunk of stone from western Greenland. The rock is metamorphic and has been dated at 3.85 billion years old. It began as sedimentary rock and was transformed by heat and compression. Sedimentary rock forms from layers of water-borne particles, which means that water has been flowing across Earth's surface for at least 3.9 billion years.

As life took hold in the oceans, oxygen levels increased in the atmosphere. But it was not a steady process. Recent research[14] shows that long after the Great Oxidation Event some 2.5 billion years ago (see the Oxygen Enigma), oxygen levels dropped again for several million years. The precipitous fall came about 1.9 billion years ago. Traces of it can be found in isotopes within ancient iron formations in Ontario, Canada. The torturous road from dust-filled hydrogen to oxygen-starved carbon dioxide finally led to the nitrogen/oxygen atmosphere we enjoy today.

14. *Nature*, September 10, 2009.

original solar nebula, the cloud of gas that ultimately formed the Sun and planets. These stony space rocks condensed from the primordial building blocks of the Solar System and are high in carbon. The second class of meteorites came from "modified" materials associated with more mature planets. As an object accrues more and more material, its gravity increases. Light elements rise to the crust and heavier ones drop to the core in a process called differentiation. Meteorites in this second class tend to be higher in metals and certain minerals.

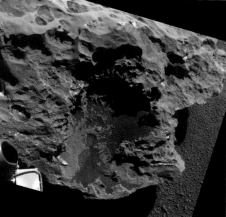

The carbon-rich carbonaceous chondrites, most ancient of the meteorites, may have contributed major amounts of Earth's carbon. It is estimated that between 37,000 and 78,000 tons of meteoritic material reaches the surface of Earth each year, with 86% as chondrites. In past epochs, the infalling rate was much greater. Up until about 3.9 billion years ago, debris left over from the Solar System's formation scattered itself across the planets, leaving scars that we see today. This "initial bombardment" is responsible for the giant impact basins that form the face on the Moon, the huge ringed plains on Mars, and the bowl-shaped impact depressions on every airless landscape from Mercury to the Moons of Neptune and beyond. After about 3.9 billion years ago, most of the cosmic leavings had been cleaned up, pulled in by the gravity of planets and moons.[15] Many of these planetary blemishes were caused by stony or metallic meteors, chunks knocked from asteroids or leftovers of that formative planetary debris. But others were excavated by the impact of comets, essentially floating icebergs. In fact, much of the early

(Left) The famous metallic meteorite Allen Hills ALH84001, found in Antarctica; (center) metallic meteorite Block Island, and (right) stony meteorite Marquette Island, both discovered on Mars by the Mars rover Opportunity (NASA)

15. Astronomers study the crater density in any given area to judge the geological age of a surface.

Comets may have played a significant part in the development and makeup of Earth's atmosphere.
Right: Comet McNaught over Aukland, New Zealand (courtesy Jamie Newman).
Left: Nucleus of comet Wild 2, Stardust spacecraft photo (courtesy NASA/JPL)

water in Earth's oceans may have been carried in by giant comets, common in the early Solar System.

The cycles of oxygen, carbon, water, and rock reinvigorate our planet, constantly replenishing its resources. Similar – and yet quite alien – systems of recycling exist on other worlds. We will find some in seemingly unlikely places.

STORMY WEATHER

If our atmosphere were able to distribute heat evenly and smoothly, every day would be just another day in paradise. But the Sun heats our world unevenly. The world's axial tilt triggers temperature differences between north and south. Shifts in currents and eddies, and differences between land and sea add to the volatility of planetary weather. Breezes become gales. Snow flurries turn into blizzards. What began as simple instabilities can give rise to the most dramatic weather on Earth: storms.

Weather disturbances come in a range of extremes. The mildest localized storms are dust devils and waterspouts. Caused by a vortex of swirling air, dust devils (or whirlwinds) are only visible when they carry sand and dirt into the sky. Unlike waterspouts, they are not rooted in clouds or strong storm systems. Dust devils most commonly occur in hot desert regions, and rarely generate internal winds higher than about 50 mph. As warmed ground heats the atmosphere, currents of warm air (called thermals) rise into clear sky. Variations in the wind speed just above the ground shove more air into the thermal, causing it to spiral upward, pulling debris with it. Dust devils also spin across the desert wastelands of Mars (see Chap. 6).

Waterspouts are stronger, and occur – as their name implies – over lakes or seas. The waterspout takes the form of a funnel of mist stretching from the base of cumulus clouds into a body of water. Like dust devils, thermals are the culprit, in this case over warm water. Although their internal winds are similar to those found in dust devils, waterspouts pose a real threat to small boats and aircraft.

The truly evil cousins of waterspouts are tornadoes. Tornadoes are the most violent and dangerous of terrestrial storms. Usually rooted in thunderstorms, they can last minutes or hours. Winds have been clocked as high as 300 mph. Some funnels can be 3 miles long and a mile across. Tornadoes generate so much energy that they can drive straw into solid wood and carry entire buildings for miles. In the aftermath of these terrifying storms, splinters of wood have been found piercing sheet metal.

One of the most powerful twisters in recorded history wound its way across Illinois and Indiana on March 18, 1925. The tornado left a trail of devastation 219 miles long. In its wake, 695 people lost their lives and over 2,000 were injured.

The center of a tornado can cause major damage, too. A tornado's interior has extremely low air pressure. Buildings sealed up and not well-vented may actually explode when the outside pressure drops during a tornado.

At times, tornadoes wander over lakes and become tornadic waterspouts. These are powerful enough to pull large amounts of water into the sky. Some witnesses have reported crayfish, frogs, and fish falling from the sky in the wake of a tornado over water.

Tornadoes are often spawned by the largest storms on our planet, the hurricanes. Hurricanes originate as tropical depressions – areas of low pressure – in a region 10° north and south of the equator, at the edges of the Hadley cells. Air moving northward or southward, away from the equator, spins as Earth turns below it. If the swirling storm grows strong enough, generating sustained winds of over 74 mph, it becomes a hurricane, also known as a typhoon or cyclone.[16]

The trouble starts over warm water. As the surface temperature of the ocean rises above 86°F, chains of thunderstorms boil up high into the troposphere. An area of low pressure pulls these storms toward its center, while the Coriolis effect of Earth's rotation gives them the curving paths that form a hurricane's notorious spiral. These lines of storms are called feeder bands. They pull heat energy from the sea and funnel it up into a massive spiral. Pressure at the surface drops further, and storms explode even higher, pulling winds from an ever-expanding area.

Like a colony of growing bacteria in a Petri dish, the storm spreads as it pulls energy from its surroundings, sucking moisture-laden warm air into its massive convective engine. Winds rush from the base of the hurricane through to the top, where they exit through a central hole called the eye. This vast pinwheel of violence can span 300 miles across, with winds up to 185 mph. The destruction of hurricanes lies not only in their size but in their duration; some storms last up to 2 weeks. Their powerful surrounding clouds generate tornadoes, severe lightning, and torrential rains. As if flooding from

16. Hurricanes form in the Caribbean, the Atlantic, the Gulf of Mexico, and the northern Pacific east of the international date line. The same storms are known as typhoons, which originate in the northern Pacific west of the date line. The storms go by the name of cyclones in the Indian Ocean.

ISS009E22187

The "eye" of Hurricane Ivan over Grenada in 2008, photographed by Expedition 9 aboard the International Space Station (NASA)

Inside the eye of Hurricane Katrina (NOAA)

rains weren't enough, piles of seawater, a "storm surge," pound the coastline in front of the oncoming maelstrom. Although hurricanes and typhoons are the most extensive storms on this planet, they are dwarfed by storms on other worlds, as we will see.

Our terrestrial ocean of air with its storm systems, water and carbon cycles, and unique blend of gases, makes us wonder about the gas envelopes cocooning faraway worlds. For centuries, astronomers were imprisoned by Earth's gravity, limited to the telescopic study of the light (spectra) coming back from the planets and moons. But the light from distant worlds cannot show us a great blue storm on Neptune, smoggy equatorial belts girding Titan, or bands of golden haze draped across Saturn. To find out the details and nuances of those alien skies, we would have to go there to ride the winds and sample the air in situ (at the source). It was time to draw plans for a field trip.

Chapter 2

Early Concepts, and What It Really Takes to Explore Alien Skies

Visionaries such as Leonardo Da Vinci dreamed of their inventions flying over Renaissance landscapes. This painting depicts Leonardo's corkscrew design for a helicopter (painting © Michael Carroll)

To the ancients, the sky was the home of the birds, insects, and bats, creatures that transcended the chains of Earth's gravity to soar above the human world. It was an unreachable realm, home to gods and goddesses. To the Hebrews and early Christians, it was the "first heaven," the place from which the rains came and the winds were born. It was the

M. Carroll, *Drifting on Alien Winds: Exploring the Skies and Weather of Other Worlds*, DOI 10.1007/978-1-4419-6917-0_2, © Springer Science+Business Media, LLC 2011

firmament of weather, and it was a place both mysterious and magnetic. People often feared the weather, but they wanted to understand the sky, too. They wanted to go there. The Greeks penned the story of Icarus, who escaped captivity on wax-fashioned feather wings. His pride brought him too close to the Sun, melting his wings and sending him plummeting into the sea. But while the story of Icarus taught and entertained, other ancients studied the details of meteorology. Aristotle wrote about the nature of freezing water and the dynamics of wind. The first attempt at classifying climate was carried out by observers in the classical Greek period. Our word meteorology comes from the Greek *meteoros*, "high above."

FIRST DREAMS

In his childhood, Leonardo da Vinci watched birds gliding above the Arno River from the Ponte Vecchio Bridge in Florence. In Milan, as he painted his famous *Virgin of the Rocks* and *Last Supper*, the artist continued to ponder the dynamics of flight. Years later, he watched the creatures of the air coast on updrafts over the canals of Venice. What could they see, far into that part of the world from which snows swirled and rainbows glowed?

It was not until past his thirtieth birthday that Leonardo began to carefully record his thoughts of flight in his sketchbooks. He designed flying machines similar to modern hang gliders. He drafted plans for a corkscrew helicopter, a scaled-up version of small flying toys of the time. He documented the workings of bird and bat wings. In one of his sketchbooks, Leonardo wrote, "The bird is an instrument functioning according to mathematical laws, and man has the power to reproduce an instrument like this with all its movements."[1]

In his Codex Atlanticus, written between 1478 and 1519, Leonardo covered over a 1,000 pages with diagrams of mechanical devices, scientific notes, and engineering studies. Within those pages, he documented the wing anatomy of several creatures, including bats, kites, and other birds.

Leonardo's most detailed flying machine was one he called "the great bird." The device was laid out so that an aeronaut would lie face down on a wooden plank beneath the wings. Using a system of pulleys, he would move the wings with his feet, pumping his legs much like the rider of a bicycle. The wing movements most closely resembled the movement of the wings of bats.

Leonardo seemed to understand, at least subconsciously, that the weight of the passenger must be low in comparison to the wings, resulting in a stable balance from a low center of gravity. His flapping machine was far too heavy to work under human power, but his studies of fixed wings bear similarity to later gliders by designers such as Otto Lilienthal and George Caley.

In another notebook Leonardo declared, "There shall be wings! If the accomplishment be not for me, 'tis for some other. The spirit cannot die; and man …shall know all and shall have wings…"

1. Codex Atlanticus, written 1478–1519.

As it turned out, the accomplishment would not be for him, and the first artificial flights would have nothing to do with wings. Instead, the earliest aerial voyages came at the hands of two Parisian papermakers. Joseph and Jacques Etienne Montgolfier noticed that burning paper – which they often saw because of their business – rose into the air. Joseph began to experiment with paper bags. Holding the open mouth of a bag over a fire, the bag became buoyant, floating to the ceiling. The Montgolfier brothers believed they had discovered a new gas, which they humbly called "Montgolfier gas." The brothers experimented with burning a variety of substances. They came to believe that various materials gave off different gases, and that those gases had different buoyancies. Though they failed to realize that simple hot air was at work, they were the first to construct a vessel that could actually make use of hot air in a practical way (politicians notwithstanding).

On the morning of September 19, 1783, a distinguished crowd assembled on the lawn outside of the Royal Palace at Versailles. Among that crowd were King Louis XVI and Queen Marie Antoinette.

A model of one of Leonardo's flying devices. Note its similarity to later designs such as those of Lilienthal (San Diego Air & Space Museum)

Early German air pioneer Otto Lilienthal flying one of his gliders, ca. 1895 (courtesy of Ursula Schafer-Simbolon, Historical Archive SDTB)

Before them, bobbing in an autumn breeze, floated an immense varnished taffeta bag some 38 ft across, filled with hot air and lined with paper. Golden tassels and signs of the zodiac emblazoned the surface of the royal blue balloon, a design influenced by the Montgolfiers' sponsor, wallpaper manufacturer Jean Baptiste Reveillon. The Montgolfier brothers wisely christened it the *Aerostat Reveillon*.

The Montgolfiers took no chances on the strength of their heated envelope. Each overlapping seam was glued together, and reinforced with a total of 1,800 metal buttons. Unfortunately for the onlookers, the inventors decided the best fuels for their first flight included a blaze of shoes and rotten meat. The King and Queen made a hasty retreat. According to one

Engraving of a Montgolfier balloon from a contemporary eighteenth-century postcard (artist unknown, Wikipedia Commons)

contemporary publication, "the noxious smell thus produced obliged them to retire at once."[2]

The hot-air balloon carried history's first air travelers. Its crew consisted of a sheep, a duck, and a rooster. The Montgolfiers considered the duck to be a control in their experiment, as it clearly could survive in the air under natural conditions. The sheep was a land creature, and constituted a good test for a mammal accustomed to living on solid ground. The chicken was a mix, a creature of the air without the natural capacity to spend time up in it.

The momentous voyage lasted roughly 8 min, attaining an altitude of about 1,500 ft. In the end, the airship touched down in a farmer's field 2 miles away, and the healthy travelers disembarked.

The royal court was impressed and entertained, but some of them must also have been envious of the animal aeronauts. The brother inventors certainly were, and set to work on a balloon that eventually carried the first humans. After several tethered tests, the brothers returned to Versailles on November 21, where they launched the first human-occupied balloon at the royal court. Jean-François Pilâtre de Rozier, a French teacher and physicist, accompanied an infantry officer named Marquis d'Arlandes into the skies above the French countryside.

In the following years, Jules Verne wrote of explorers who circled the world by balloon in his popular *Around the World in Eighty Days*. But Verne dreamed of skies even more distant. And while he wrote of men traveling to the Moon, his *From the Earth to the Moon* found its publication seventeen centuries late, long after another writer envisioned such a trip. In AD 160, Greek satirist Lucian of Samosata wrote of Menippus, a man carried to the Moon by a waterspout.[3] Over a millennium later, German mathematician/astronomer Johannes Kepler wrote the *Somnium* ("The Dream"), the adventure of a young student whisked off to the Moon by lunar demons during a solar eclipse. Kepler used the seventeenth-century story to scientifically describe the appearance of Earth from the Moon, and to defend the Copernican (Sun-centered) view of the Solar System. Even J. R. R. Tolkien[4] wrote *Roverandom* (1928), the story of a toy dog's adventures on the Moon.

THE VIEW FROM AFAR

With the invention of the telescope, the planets became a serious topic of scientific inquiry. Observers in the eighteenth and nineteenth century began to wonder: do these worlds bear similarity to our own? Do they have days

2. From *Memoires secrets pour servir a l'historie de la republique des letters en France dequis 1762*, quoted in James Glaisher, Camille Flamarion, W. De Fonvielle, and Gaston Tissandier, *Travels in the Air* (J.B. Lippincott, 1871, p. 59).

3. Menippus's journey is described in Lucian's *True History*.

4. Author of the *Lord of the Rings* trilogy.

and nights, summers and winters, clouds and wind and rain? Venus seemed to be covered in clouds, and some observers thought that the distinctive brightening events on Martian plains were caused by dust storms. Jupiter certainly seemed to be a cloudy world, with its colorful streamers, ovals, and bands.

It would take centuries of focused effort to unravel the tangle of mysteries associated with weather on other worlds. Early researchers considered it likely that the closest planets, Mars and Venus, were quite earthlike. Many believed Venus was "younger" in evolution than Earth, perhaps imbued with an environment similar to Earth's in the Carboniferous period. Mars was regarded as an ancient world, with an atmosphere thinner than ours, perhaps even a dying planet. Percival Lowell (1855–1916), an amateur astronomer who popularized the idea that Mars was an "abode" of intelligent life, said, "A planet may in a very real sense be said to have a life of its own… It is born, has its fiery youth, sobers into middle age, and just before this happens brings forth, if it be going to do so at all, the creatures on its surface which are, in a sense, its offspring…in the special case of Mars, we have before us the spectacle of a world relatively well on in years, a world much older than Earth."[5]

And what of the gas giants – Jupiter, Saturn, Uranus, and Neptune? Through the telescope, the fuzzy planets beyond Mars seemed even more mysterious. Jupiter, closest and largest, displayed alternating light and dark

An 1865 illustration of Verne's explorers getting their first experience with the weightlessness of translunar space. From his novel From the Earth to the Moon

5. *Mars* by Percival Lowell. Originally published by Houghton Mifflin in 1895, special limited reprint 1978, pp. 206–207.

Strange Parallels

Space exploration cannot happen without the visionaries, engineers, and scientists first dreaming of what might be. French writer Jules Verne is considered one of the most important futurist writers of the nineteenth century, and one of the founders of the science fiction genre. His novel *From the Earth to the Moon* is a typical example of Verne's carefully thought out science and prescient hunches. *From the Earth to the Moon* describes the first lunar voyage. The story is rife with parallels to the first real Moon mission, the *Apollo 8* flight in 1968. Here are some of those parallels:

Jules Verne's Mooonship carried a crew of three.
The Apollos carried crews of three.
Verne's ship was shot from a cannon in Florida.
Apollo 8 lifted off at a Florida site just 213 km north of Verne's site.
Verne's cannon shell was made of aluminum and weighed 19,250 lb.
The Apollos were constructed of aluminum and weighed 26,275 lb.
After circling the Moon several times, Verne's ship returned home to a landing in the Pacific Ocean, where it was recovered by a Coast Guard ship.
Apollo 8 splashed down in the Pacific Ocean, where it was recovered by a Coast Guard ship.

bands arranged parallel to its equator. As telescopes improved, nineteenth-century observers could resolve details such as streaks, ovals, swirls, and multicolored festoons that came and went, changing color over time. Saturn and Neptune both hosted their own faint versions of those bands. What were astronomers seeing? What did it mean about weather on these behemoth worlds, and what was below? Estimates ranged from overcast swamp-worlds to vast volcanic wastelands enshrouded in chilled, poisoned gases. The planets were numbingly remote. How could we discover their natures from such a distance?

A WIDE SPECTRUM OF INSIGHTS

In 1835, the French philosopher Auguste Comte felt this same frustration.[6] On the subject of stars, he wrote, "While we can conceive of the possibility of determining their shapes, their sizes, and their motions, we shall never be able by any means to study their chemical composition or their mineralogical structure … or even their density… I regard any notion concerning the true mean temperature of the various stars as forever denied to us." A scant 14 years later, German physicist Gustave Kirchhoff discovered that the chemical composition of gas could be determined by splitting the light that it emitted into a spectrum. The new science of spectroscopy was born, and with it, astronomers had a new and powerful tool.

Telescopes began to reveal detailed information with the advent of spectroscopy. Spectroscopy, the study of all the colors in light, gave observers the first information about the constituents in those distant atmospheres. Light from distant planets held wonderful treasures, fingerprints of their components. Dark patterns called absorption lines appeared in unique patterns across the spectrum, revealing what materials were present in the objects being studied. This breakthrough enabled astronomers to send light from their telescopes through a prism. The resulting rainbow of colors enabled observers to chart the dark absorption lines from any planet or moon they wished to study. Even from millions of miles away, materials left traces on the light reflecting off the surfaces and atmospheres of distant worlds.

"Spectroscopy is one of the planetary scientist's favorite tools because you can study atmospheres from a distance," says Nick Schneider, researcher at the University of Colorado, Boulder. Schneider's studies of Jupiter and Io rely on spectroscopy. Using the light signature from these worlds, Schneider can figure out if he is looking at gas rather than ice, and even gives insight into how much atmosphere is there. "It's a subtle thing. A solid vs. a gas affects the line's shape. A gas has the purest spectrum. Individual molecules emit and absorb at these characteristic wavelengths very precisely. But if they are bumping into each other or distorting the shape of the molecule, if you have them in a crystal where one molecule can nudge adjacent molecules, all those absorption and emission bands get fuzzed out a little." Astronomers know that not only does the band's shape change, but also even the exact

6. Comte was writing in
*Cours de la Philosophie
Positive*, 1835.

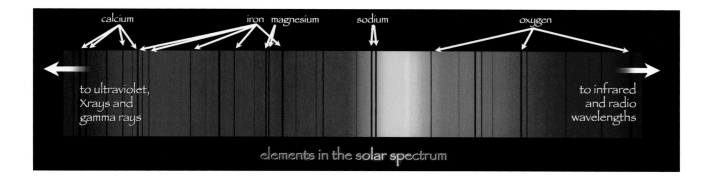

calcium   iron magnesium   sodium   oxygen

to ultraviolet,
Xrays and
gamma rays

to infrared
and radio
wavelengths

elements in the solar spectrum

Elements within the Sun – simplified here – leave telltale tracks in a spectrum of sunlight (© Michael Carroll)

wavelength where most of the absorption happens shifts a little. "We know this because people love to take tanks of gas and cool them down to really low temperatures, shine a light through, and see what the spectrum looks like, or freeze that gas onto a surface, shine a light on it, and see what comes back. This is all confirmed to many decimal places in the laboratory."

Spectroscopy also yields information about the speed at which an atmosphere is escaping into space, says Schneider. "I study sodium escaping from Io. It emits at the same two wavelengths as the color from yellow–orange streetlights. So the first thing I do is say, 'Oh, here are those sodium emissions. I know how much sodium is there.' But once I have that fingerprint of two colors of yellow–orange light at this exact spacing apart, I can use their Doppler shift to figure out if the sodium is coming or going relative to Io. I can measure how rapidly Io's atmosphere is escaping." In effect, scientists use a spectrograph as a speedometer.

Another tool at hand for astronomers is that of stellar occultation. MIT's Heidi Hammel describes the technique as it was used to probe the atmosphere of Neptune's moon Triton. "You watch a star, and as Triton moves in front of the star, instead of the starlight being blocked out instantaneously by the rocky surface, it drops off gradually and then disappears. That gradual dropoff is a measurement of the properties of the atmosphere surrounding Triton. By carefully modeling the curve as it drops off, you can infer the pressure and temperature, assuming you know the composition."

Although telescopes provided insight into what constituents floated in alien skies, the instruments afforded only limited detail about the structures of those atmospheres. Were there clouds? Fog? Lightning? Swirling storms? Clear, deep atmospheres? Were other worlds like our own, waiting for us to set up camp?

The Space Age changed all of that. As the data from the first spacecraft trickled back to Earth from Mars and Venus, it became more obvious that there is no place like home. Early data from robotic spacecraft began to paint a more dismal picture of even the worlds close by. A 1964 episode of the popular *Twilight Zone* television series reflects the more pessimistic view science was beginning to adopt. In "The Long Morrow,"[7] a scientist tries to convince an astronaut to travel to the nearest star because of the bleak

7. *The Long Morrow*, written by Rod Serling, © CBS, first broadcast on January 10, 1964.

*Atmospheres of the planets
and moons display a range
of structure and depth.
Left to right: Thin hazes over
Mars, Earth's "medium"
atmosphere, Titan's dense
orange gases, and the limb of
the gas giant Saturn (NASA)*

Atmospheres of the planets and moons display a range of structure and depth. Left to right: Thin hazes over Mars, Earth's "medium" atmosphere, Titan's dense orange gases, and the limb of the gas giant Saturn (NASA)

outlook in our own Solar System. The character says, "And what do we know about our neighbors, Commander? Mars is a vast, scrubby desert with an unbreathable atmosphere. Pluto is poisonous and extremely cold. The Moon is barren, Jupiter volcanic. In short, Commander, our neighbors offer us only one asset: they are accessible…beyond that, they offer us nothing scientific, social, economic, anything."

This killjoy scientist may have spoken a bit prematurely. Today, we see a menagerie of planetary atmospheres, ranging from diffuse to crushing. Although the worlds of our Solar System possess clearly hostile environments, many hold the promise of untapped resources and places well worth exploring. But how do we get there?

GETTING FROM POINT "A" TO POINT "B"

Waterspouts and lunar demons notwithstanding, the only way to get to another world with today's technology is by brute force.[8] Sending a craft to distant planet is like throwing a rock. All the force is unleashed at the start, so the path must be true to begin with. The power involved in casting a payload across the void is remarkable. Just getting to another planet turns out to be a violent and terrifying prospect.

Rockets, the choice of planetary exploration so far, have been around for a long time. The earliest solid rockets were nothing more than glorified fireworks. Called "fire arrows" by their first-millennium Chinese inventors, they were aimed in the correct general direction and let fly with fingers crossed. Fins added stability. Much later designs incorporated liquid fuel, which has a much more powerful impulse (thrust during a given time) than solid rocket propellant. The more advanced launchers were far more complex, and this caused problems in early exploration. Says Lockheed/Martin's Bryce Cox, "They'd get 4 ft off the pad and just explode, or 10 ft off the pad and they'd auger over and blow up."

Even today's sophisticated launch vehicles face daunting challenges and subject their delicate robotic explorers to remarkable stresses. Planetary spacecraft must face the temperature extremes of space, the stresses of

8. An exception to this rule is solar ion propulsion, which uses solar energy to ionize a heavy gas such as xenon. The gas is then magnetically charged and accelerated by magnets to create a weak but steady thrust.

vacuum, and long cruise periods. But the beginning of their journey is just as dangerous.

Before launch, the delicate payload remains in an air-conditioned, carefully monitored environment. The fairing, or shroud, in which it rests atop the booster is pressurized slightly so that no debris can make its way in from the outside. Instrumentation tells ground controllers about the health of both booster and spacecraft, a luxury that early space missions did not have.

Once the launch vehicle is declared healthy, and the celestial realm is aligned for a launch opportunity, the rocket engines ignite. "You hit T-zero – ignition – and often the motor start-up can be one of the bigger shocks to the payload," says Cox. "It can be quite a kick."

The thunderous ignition reverberates, shaking windows miles away. This sonic tidal wave can damage equipment and booster structures. To dampen it, the launch pad is equipped with a water deluge system. Gushing water helps to soften the acoustic energy pouring from the rocket's engines. As the vehicle ascends, the sound no longer bounces back from the launch pad, but it continues to buffet the very core of the rocket and its cargo.

America's first attempt at orbiting a satellite in 1957 was a spectacular disaster. The grapefruit-sized satellite was the Vanguard 1 (NASA)

Large surface areas of a payload are most susceptible to vibration. Equipment such as antennas and solar panels are clamped securely to the body of the spacecraft, often with padding to further protect them. The interior of the payload area may also be blanketed with sound-suppressing material and can even be filled with lighter-than-air gas that transmits sound more poorly than the external environment.

Once airborne, the booster pitches over into the correct flight path. Every rocket has its own frequency, expanding and contracting and flexing as it travels. Each turn and trajectory change causes it to shift like a living, breathing creature. Cox likens an ascending booster to a great undulating coil. "The launcher is like a big spring with a mass hanging off the top – the payload – plus all the other stages down below acting like more springs stacked on each other. The whole thing can get wobbly. You have to know what those fundamental frequencies (the amount of flex) of the launch vehicle are, and then you try to design so that all your thrust controllers and guidance algorithms are separated from those modes." One tool at the designers' disposal is called dual plane isolation. Between each stage, a donut-shaped shock absorber softens flight stresses. These isolators filter out vibration and shock.

As the speed of the craft increases, the air becomes an enemy. Dropping air pressure outside necessitates venting of the air inside the shroud. But the air pressure against the front of the vehicle increases as air rushes against it, heating up the protective shroud at the vehicle's nose. Temperatures can rise more than 350°F. Some materials such as aluminum begin to lose their structural integrity as they heat up. This degrading, called thermal knockdown, comes at just the wrong time, when the vehicle is reaching its maximum stress. This period is called Maximum Dynamic Pressure (or Max Q). Many flight profiles call for reduced power during this critical time.

The shape of the nosecone can help a booster through its Max Q stage. Aerodynamically efficient shapes plow through the atmosphere with less stress than do wider payloads. Wider shrouds must be stronger to survive supersonic, then hypersonic, voyages through the atmosphere. "Think of it this way," Bryce Cox explains. "If you jump off a diving board and go in the water fairly straight, it doesn't hurt too much, but if you jump off and do a belly flop your velocity, for all intents and purposes, is nearly the same, and the density of the water is the same, but you have a lot more surface area exposed; thus the sting! The dynamic pressure is the same but the results are different. For rockets this 'difference' gets translated into structural design capability. The more surface area exposed, the heavier it has to be to be structurally capable."

As the booster and spacecraft continue into higher levels of the atmosphere, the first stage is dropped and the second fires. Along the way, the shroud is jettisoned, as it is no longer needed in the thinning air. A third stage usually follows the second. Once in orbit, the payload must survive yet another shock as it blasts onto a path that will take it to its ultimate planetary destination.

BUILDING A BETTER PLANET-TRAP

Aside from arriving in once piece, successful planetary missions were challenged by a simple lack of scientific knowledge. Early Soviet designs for Mars probes assumed a much denser atmosphere than the one we now know exists. Had their first Cosmos probes made it to the atmosphere, they would have perished under parachutes far too small to do the job. Soviet Mars and Venus probes were initially patterned after designs for the Luna program, which successfully landed on the Moon first in 1966.[9] But as knowledge of Venus and Mars accrued, engineers realized that more robust craft were needed to visit alien atmospheres. Thus, designers drew their plans for the larger and more sophisticated Venera (Venus) and MAPC (Mars) probes.

Like the Soviet Union, the United States had a burning interest in seeing what was in the clouds above the two nearest worlds. One of the engineers who worked on early planetary designs was Patrick Carroll, an aerodynamicist at Martin Marietta Corporation. "There were a lot of unknowns," Carroll remembers. "Visually, you couldn't see anything on Venus from here. It was tough to get good design data. That was true even with Viking at Mars;

9. The Soviet *Luna 9* touched down on the lunar plain called the Ocean of Storms on February 3, 1966.

we didn't know if we'd sink in up to the deck or not. It added an interesting aspect to design."

With little data to go on, scientists and engineers combined forces to try to build a probe that just might survive entry into an alien sky. As a space probe approaches any planet, it gains speed under the influence of the planet's gravity. At the point that the probe reaches the upper fringes of the atmosphere, it is usually traveling at a speed at least equivalent to the planet's escape velocity. At the time the first probes were being considered, no craft had survived an entry at such high velocity or temperatures. The proposition was intimidating. American engineers came up with a conical design for an ablative heat shield, a protective cover that would gradually burn off as the craft slowed. The flat cone provided aerodynamic stability, while being blunt enough to eventually slow an incoming craft so its parachutes could deploy. But the harrowing speeds and unknown details of atmospheric structure made designers nervous. Today, computers can model an entry profile, telling designers whether their proposed craft will survive entry or not. In the early 1960s, only the simplest of models existed, done not by computer but by paper and human creativity. Carroll's department was no exception. "Each area had one machine that was a crude computer, much like a cash register. There were punch cards and slide rules. We made do and worked on Saturdays."

Soviet engineers came up with a different design. Like a Russian matryoshka doll, their planetary robots nested inside a metallic globe. The sphere was weighted so that one side would tend to point in the direction of travel. Soviet designs also called for an ablative material to burn off as the sphere entered the atmosphere. The Soviet design worked well, but was unknown to the west for some time. The Cold War was in full swing, explains Carroll. "There was essentially no communications with the Russians, no interface except what we – or they – could steal from each other."

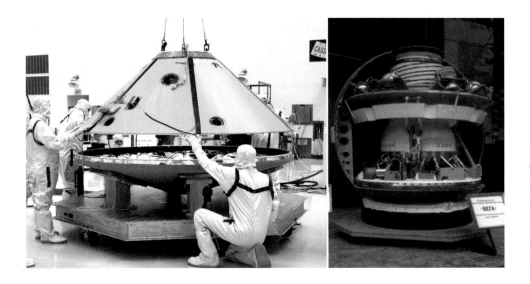

(Left) U.S. spacecraft entered alien skies using a conical heat shield, like this one on the Mars Exploration rovers (NASA/JPL); Soviet era and Russian planetary probes have spherical heat shields, as in this model of the Vega Venus lander (© Ron Miller)

So the Soviet and U.S. programs continued on in their separate ways, with stops and starts, failures and successes. With successes of early Moon probes, scientists felt encouraged that the time would soon come for a successful exploration of planetary atmospheres. But once a probe was safely delivered to the atmosphere, what might it do? For early researchers, the sky was indeed the limit. From simple probes descending slowly on parachutes, designers' techno-dreams expanded to balsa airplane robots plying the skies of Mars and Venus. Some envisioned dirigibles tacking against alien winds for months at a time, relaying information to orbiters overhead. In a cosmic bow to Verne's *Around the World in Eighty Days*, still others preferred Montgolfier-style open-air balloons to explore Venusian clouds and Martian mists. Biophysicist Ben Clark joined planetary exploration with the Viking Mars landers in 1975. He learned of NASA's history and heritage as he walked the halls of NASA's Jet Propulsion Laboratory and later, Lockheed/Martin. "The atmosphere [of Mars] was originally thought to be 80 millibars [ten times what it really is]. We were designing Mars airplanes back then." Those stubby-winged designs were overly optimistic.

Remote sensing refined estimates of the atmospheres of Venus and Mars, forcing a shift in approaches. Venus simmered under copious blankets of dense gases, making entry – and eventual landing – relatively easy in its dense air. But Mars was a different story. Telescopic observations of star occultations, measurements of a star's dimming light as it passes behind the atmosphere as viewed from Earth, showed the Martian air to be dangerously thin, perhaps thinner than that found on Everest. Engineers such as Ben Clark began to wonder if it was even practical to land in the rarefied Mars air. "It turns out that Mars is the hardest place to land on in the Solar System. The atmosphere is thick enough that you can't do a propulsive landing like you can on the Moon or with asteroids. But it's not thick enough to just land on parachutes like you do on Earth. You have to have an entry capsule, then a parachute to slow you down, and then you've got to get out of that aeroshell to activate your airbags or fire up descent engines. Mars is very tricky that way."

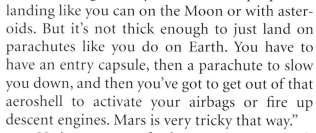

Various spacecraft planners came up with scenarios for skimming along the outer layers of atmosphere to slow down, plunging in with armored probes, or decreasing speed with clever inventions. One such invention was Pat Carroll's brainchild – a hypersonic drag device. The metallic parachute would spring open while the craft was still traveling far too fast for conventional parachutes, and would be jettisoned, red-hot, after delivering the falling probe to the denser region of the atmosphere. There, a normal parachute could open safely. The idea would be revisited decades later as designers laid plans for the first atmospheric visitor to the outer planets, the Galileo Jupiter probe.

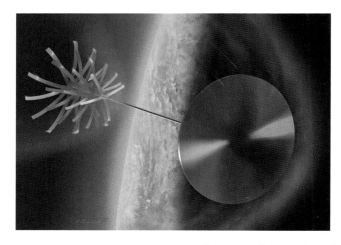

This concept for a hypersonic drag device, patented in 1966, called for umbrella-like strips of metal to slow a probe in the upper atmosphere of a planet (© Michael Carroll)

Venus inspired more designs. Because of its dense atmosphere, early plans included a balloon that would inflate to keep a probe up in the cool regions of Venus's hothouse weather. The concept was a complex and expensive one, especially for early 1960s technology, but its day would eventually come.

In the meanwhile, aerospace engineers considered what data could be collected by a descending probe, whether on Venus or Neptune. How much science could you do in the brief time that a parachuting spacecraft had? How long could a probe survive? How could you get scientific data back to Earth from a tiny craft swinging on parachute or balloon lines in alien winds? Early designs showed probes brimming with Rube Goldbergian devices. Scoops, vents, and tubes protruded at wild angles from conical probe bodies. Martin Marietta's Carroll was inspired to draw a cartoon depicting the spirit of his team's early plans. "They wanted to get all kinds of chemical measurements and heat profiles as you went in. Practically every scientist with a different specialty was interested in putting their experiment on it. You have to balance that against how much you can take. There was a lot of uncertainty as to what we were going to get once we were there. You'd be measuring stuff all the way down."

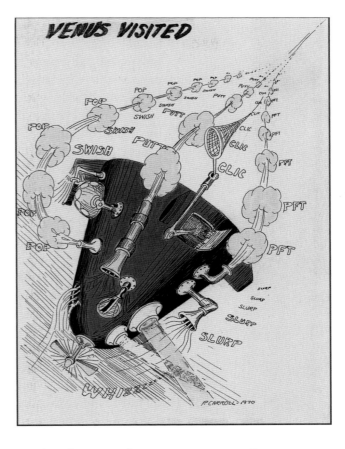

Patrick Carroll's cartoon spoofing a Venus probe study. The actual design called for even more devices than his drawing depicted (© Patrick Carroll)

ARRIVAL

Every planet confronts explorers with a unique set of challenges because of the unique nature of its atmosphere. But all worlds share some common hazards to entering spacecraft. After a harrowing launch with vibrations that can break wires or damage instruments, interplanetary robots must brave months in the hard vacuum of space. But more stresses are to come. Aside from launch, the meteoric entry into an atmosphere is the most dangerous part of an atmospheric probe's flight. As it hits the upper fringes of air, friction envelops the spacecraft in an incandescent ball of fire. The heat of entry is more than enough to destroy the spacecraft if it were to absorb all of it. After all, meteors of stone and metal regularly burn up in our own atmosphere. For a robot explorer's survival, something must be done to dissipate the heat.

Initially, molecules of rarefied air bombard the spacecraft. The air flows freely around the craft, rushing across its face while missing sheltered parts of the structure. But as the craft descends into denser gases, air in front of the craft stacks up and heats just ahead of the craft in a shock wave. Like water before a speedboat, this superheated wave rides in front of the craft. How close these searing gases come to the spacecraft depends on the shape of the vehicle. Pointed, slender objects move through the atmosphere with a shock wave propagating directly from their point, streaming back in fairly straight lines. If the object is flatter, as in a blunt cone, that shock wave is dispersed and keeps its distance from the surface of the craft. An oblate shape seems the easy way out, but there is a problem: instability. If the protective cone – called an aeroshell – is too shallow, it will wobble and eventually tumble, breaking up in the hypersonic airflow around it. The center-of-mass of the precious cargo inside the entry shell must be forward far enough to keep it stable, and this places the delicate probe nearer the heated surface of the aeroshell.

Heat flowing around the craft, even if it is separated by a shock wave, is usually enough to severely damage whatever is inside. To get around this problem, all planetary explorers have been covered in ablating materials that gradually burn off during entry. As they vaporize, they also reduce heat by flowing into the gap between the shockwave and aeroshell, creating another boundary between the heated gases and the payload.

In many cases, the most unstable time in flight is during the transition from supersonic to subsonic flight. Because of this, many probes are spun up and/or release a small drogue parachute to keep them stable during this phase. The deployment of parachutes or various instruments can be triggered by a "g switch," a spring-loaded accelerometer that senses the probe's change in speed. But early researchers realized that accelerometers could also be used for science. The rate at which a spacecraft slows in an atmosphere gives clues to the changing density of the air outside. Temperature and pressure can also be estimated with a great deal of accuracy using data from accelerometers.

Once the spacecraft is safely past the highest heat and deceleration, a parachute is usually deployed to slow its descent and to free the probe from its aeroshell. The flow of air during descent can be used to force air samples into the craft for testing, and vanes can be placed around the probe to create spin for scanning instruments such as photometers (light sensors) and scanning imaging systems. Wind vanes were used on both the Huygens-Titan and Galileo-Jupiter probes.

Sometimes a probe needs to descend faster than a parachute allows. If it is falling through a thick or deep atmosphere, communications may dictate that it make it to a certain depth in a given amount of time. This is especially true if its signals are relayed to Earth by a circling orbiter or passing mother craft. Several Venus probes jettisoned their parachutes at relatively high altitudes to get to lower levels before succumbing to the surrounding heat. The Huygens-Titan probe released its larger parachute to speed its descent. Flight engineers often design free-fall into some portions of flight for probes in the outer Solar System to access deeper levels of atmosphere in a reasonable amount of time.

DEEP SPACE

More advanced concepts were needed for destinations beyond Mars. The outer planets – massive spheres of hydrogen-rich gases – orbit in the frigid realm beyond the Asteroid Belt, where air is chilled to mind-numbing temperatures, liquid gas falls as rain, and lightning crackles in miles-long incandescent streams. With Jupiter, Saturn, Uranus, and Neptune, distance was the first big challenge. No probe had survived travel to the nearest of worlds, let alone a years-long journey to Jupiter, nearest of the four gas giants. Thus, Jupiter became the focus for the handful of outer planets studies in the 1960s and early 1970s.

As early as the late 1960s, engineers envisioned spacecraft equipped to explore the gloomy outer reaches of the Solar System. Solar power was out of the question; distances from the Sun and the power required to get signals back to Earth precluded its use. Batteries were of no use on a decade-long cruise. The only energy adequate to the task was nuclear power. The first focused study, carried out by the Jet Propulsion Laboratory, took the form of TOPS, the nuclear-powered Thermoelectric Outer Planets Spacecraft. The TOPS design was dominated by a 14-foot diameter antenna. The craft would have been powered by four radioisotope thermoelectric generators, or RTGs (plutonium power sources). As it flew by each planet, the mother craft was designed to drop off atmospheric probes. Its launch weight was projected at over a ton (1,446 lb).

Distance, asteroids, and radiation all confronted the survival of the TOPS craft. But the biggest danger, the one that killed the Grand Tour even before its launch, was government budget cuts. As originally envisioned, the multiprobe TOPS was simply too expensive. NASA would need to come up with a different plan for a flyby craft, but in the meanwhile, engineers pondered the design of an atmospheric probe for the outer planets.

The Thermoelectric Outer Planets Spacecraft, first proposed to take the "Grand Tour" of the outer planets, was projected to be too expensive. Painting by Chuck Bennett (© Lockheed/Martin)

The biggest problem for a Jupiter atmospheric probe was speed. A typical entry profile has a probe entering the Jovian atmosphere at a speed of over 100,000 miles per hour. Some designers assumed that survival after entry was not possible, so they were tasked with designing a Jupiter turbopause probe, a probe that would return data from the uppermost atmosphere just before burning up. More daring plans called for a probe to survive the explosive entry and explore Jupiter's multicolored cloud decks as it swung beneath a parachute or balloon.

"Scientist groups would have big roundtables trying to figure out

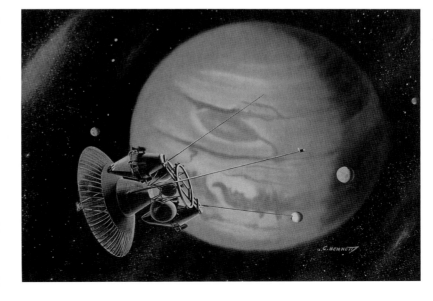

Jupiter turbopause probe, by Chuck Bennett (courtesy Lockheed/Martin)

Soviet design for a flyby/ probe mission to Jupiter. This classic painting was done by Soviet artist Andre Sokolov

what would be the right goals, what to measure, and that sort of thing," Carroll says. "You generally couldn't do it all with one probe. We had to create different designs. [The scientists] were constantly discussing what to put on it. Every meeting we had, there were major changes and different emphasis. It was fun. Everybody was very cooperative. They kicked all this stuff around, and the group would compromise. They kept getting smarter with what they could do from here, with telescopes, etc., and when they learned something new it changed something for us. Things were constantly fluctuating. That's the way you do engineering. There are advances in technology as well as advances in knowledge about the target. You want to be able to adjust your programs right up until they launch it off. Then the discussion's over. It's on its way, and you take what you can get."

Soviet scientists drew their own plans. One proposal envisioned a flyby craft that could deploy an atmospheric probe. The probe would descend without a parachute in order to remain in contact with the flyby vehicle for as long as possible.

Preliminary U.S. studies were also done of atmospheric probes for Saturn and Uranus, though these were mere formalized ideas compared to Jupiter and Titan studies of the early 1970s. While some visionaries dreamed of Jovian cloudscapes and Saturnian sunsets, others looked to the moons of the gas giants. Moons in the outer Solar System range from mountain-sized rocks to nearly planet-sized spheres of stone and ice. Ice reigns in the realm of the gas giants; most moons are either covered with it or made of it. The four largest moons of Jupiter are members of this exclusive chilly club. Discovered in 1610 by Galileo Galilei, they are called, cleverly enough, the Galilean satellites. The smallest, Europa, is about the size of Earth's Moon. The largest, Ganymede, is larger than the planet Mercury. None of the Galileans have conventional atmospheres, although rarified oxygen and hydrogen have been detected at Ganymede and Europa. Io floats in a vast cloud of sodium belched

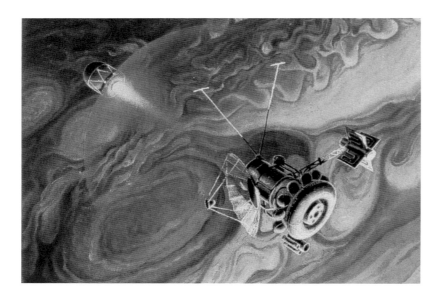

from the throats of its hundreds of volcanoes. As a rule, no substantial atmospheres swath the 164 moons known in the outer system at Jupiter, Saturn, Uranus, and Neptune. The exception is Titan, the mysterious, planet-sized satellite orbiting Saturn.

Titan baffled astronomers from the moment they were able to resolve its spectrum in rudimentary telescopes. It was huge, reddish, and clearly had some kind of atmosphere. Christian Huygens first spied the moon in a telescope of his own design during the spring of 1655. Inspired by Galileo's work, Huygens had set out to find moons of other planets, and his search paid off at Saturn. Over the next few decades, observers calculated Titan's orbit around Saturn, and its approximate size and brightness, but not much more progress could be made for three centuries. At the beginning of the twentieth century, Spanish astronomer J. Comas Sola observed that the edge of Titan faded to a darker color than its ruddy center. This phenomenon is called limb darkening, and is common in planets with dense atmospheres. Sola suggested that Titan had such an atmosphere. He was right. The proof came through the work of Gerard Kuiper in the 1940s. Kuiper was doing a systematic study of the spectra of planets and moons. When he came to Titan, he discovered that the massive satellite reflected light consistent with methane. But Titan's spectrum could not tell him how much air it had. Estimates ranged from a thin, wispy atmosphere to a dense fog like that of Venus.

Less than three decades later, with little more information that what Kuiper had gleaned, engineers were already planning missions to Titan. Saturn is a distant object, orbiting nine times as far from the Sun as Earth

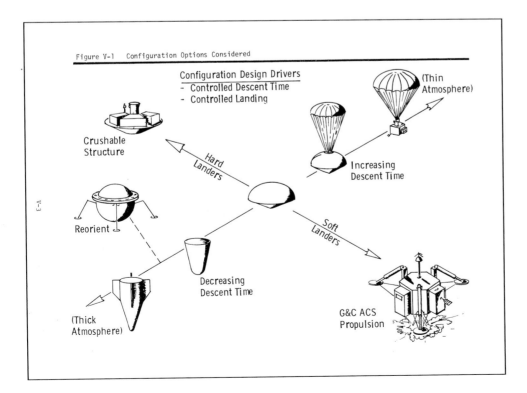

When the first studies of Titan probes were commissioned, scientists were uncertain about the density of the atmosphere. This diagram demonstrates the engineering challenges brought by the wide spectrum of possibilities considered for Titan's environment (From the collection of P. Carroll)

A probe drops a flare over the dark landscape of Saturn's moon Titan in this 1976 NASA/Ames/Martin Marietta technical study (© Michael Carroll)

does. A trip to Titan was at the edge of technological capability when, in the summer of 1976, several aerospace companies began to seriously study a robotic voyage to the frozen, dark world.

Pat Carroll was on Martin Marietta's team. "In our studies we had to design for 'what if s' of all sorts that could wipe you out, like landing in liquid. This study was done before even Voyager, so we had no idea what was down there and what conditions we would be flying through. You had to design for all those contingencies in those early studies. Of course, you didn't always have a very good solution, but you worked with the best technology and knowledge of the time."

Scientists knew that the temperature of Titan's surface is at the triple point of methane. That is, surface conditions could support methane in a liquid, solid, or gas form.[10] This meant that any atmospheric probe might encounter cryogenic drizzle, fog, buffeting winds, or even lightning in those alien orange skies. Carroll's team had to take the possibilities into consideration for any design of a lander. "We were prepared to land either in the soup – whatever it was – or a snow bank, or a solid surface. We were just trying to design the mechanics of surviving all these possibilities. We didn't have much detail. It looked complicated. The mission was a chancy one."

The uncertainty about conditions on Titan is reflected in a February 1975 technical proposal done for NASA Ames by Martin Marietta:

> Polarization measurements by Veverka and Zellner, combined with Titan's low UV albedo, suggested the presence of optically thick clouds and raised the possibility of an extensive and spectroscopically unobserved atmosphere beneath them…the surface temperature could be as high as several hundred degrees Kelvin and the surface pressure as great as several bars. However, the opposite extreme was also shown plausible when Danielson, Caldwell, and Larach demonstrated that the anomalous infrared measurements could be interpreted as a temperature inversion in a thin atmosphere…the measurements of Titan's radius are difficult and the earlier results may have been inaccurate…[11]

Scientific ignorance about Titan's atmospheric structure led to yet another problem – imaging. What were the light levels like? At Titan's distance, the Sun shines with a feeble one percent of what it does on Earth, and evidence suggested that at least some cloud cover could effectively block sunlight from reaching the surface, dropping noontime light levels to a dim twilight. Light amplification devices of the 1970s were impossibly energy hungry for a probe, and CCD technology was in its infancy. One clever proposal called for the probe to jettison a flare. The artificial light source would drop

10. Earth, for example, is at the triple point of water.

11. *A Titan Exploration Study – Science, Technology, and Mission Planning Options,* Martin Marietta report P75-44480-1, February 1975.

Two Early Titan studies: Left: A 1975 study of a Titan rover. The main body and spoked feet would deploy from a crushable landing sphere. The craft was to carry imaging equipment, soil samplers, and a meteorology station. Right: A lander deploys a helium balloon that carries a camera and meteorology station. The balloon's height can be adjusted for more distant views and sampling of higher layers of atmosphere (sketches © Michael Carroll, based on diagrams from Martin Marietta technical report P75-44480-1)

With scientists unsure of light levels on Venus, Veneras 9 and 10 carried floodlights, seen here mounted on the landing struts (NASA; NSSDC catalog/ GSFC/NASA)

beneath the probe to illuminate close-in atmospheric structure and distant surface features.

Decades later, the European Space Agency's Huygens-Titan probe carried a light source to illuminate its immediate surroundings and was equipped with CCD imaging systems that could operate in low light levels.

Titan was not the only target for which artificial lighting was carried. Early Soviet Veneras carried small floodlights. In both the case of Titan and Venus, light levels turned out to be significantly higher than feared.

The reality of exploration has always followed the vision of a few. From Leonardo and Lilienthal to Carroll and Clark, dreamers led engineers to blaze trails that would ultimately lead to distant worlds. But the first glimpses we would have of planetary meteorology came not from within planetary atmospheres but rather from above them. Our earliest explorations of the skies over planets and moons came to us through the eyes of orbiters or spacecraft that simply flew by.

After a fourteen-year mission, flight engineers commanded the Galileo Jupiter orbiter to self-destruct in the atmosphere of Jupiter, protecting Europa and other moons from possible future contamination by the retired spacecraft (© Michael Carroll)

Chapter 3

Studies on the Fly

It could be argued that any planet or moon of significant size has an "atmosphere." For example, a mist of oxygen atoms drifts along the surface of Jupiter's moons Europa and Ganymede. Those oxygen atoms are the product of solar and Jovian radiation blasting oxygen from their water-ice surfaces. Other moons belch transient atmospheres into their black skies. Geysers on Saturn's Enceladus toss 150 kg of water into the Saturnian system every second. The volcanic moon Io cocoons itself within a cloud of erupted sulfur, which it drags – comet-like – behind as it circles Jupiter. Out at Neptune, the moon Triton lets fly with its own geysers, columns of super-chilled nitrogen.

However, all of these atmospheres are so rarefied that they cause little of the weather-related phenomena we know on Earth. There are seven worlds beyond our own that host extensive atmospheres. These are Venus, Mars, Jupiter, Saturn and its planet-sized moon Titan, Uranus, and Neptune. Here, we will find weather both familiar and truly alien.

A final destination, Neptune's moon Triton, lies somewhere on the brink between weather and vacuum. Triton has a thin, vaporous atmosphere, but temperatures and cryovolcanism contribute to a unique environment that is worth visiting. The Solar System offers cosmic meteorologists a smorgasbord of weather and climate. As we work our way outward from the Sun, we will see patterns of change along with a few surprises.

VENUS: HOTHOUSE UNDER THE FOG

America's first attempt to study the atmosphere of another world at close range commenced on July 22, 1962. The Mariner 1 mission lasted 293 s, and ended in flame and frustration. The Venus-bound booster veered off course shortly before the first stage was to drop off and had to be destroyed by the Range Safety Officer. NASA studies showed the failure resulted from the combination of loss of lock on ground guidance data and a software failure.[1] Software and guidance experts scrambled to decipher the cause of the failure. A sister craft awaited imminent launch.

Mariner 2 became the first successful interplanetary probe, launching exactly 1 month later. Mariners 1 and 2 were based on the design of the successful Ranger Moon spacecraft. They stood 9 ft tall and sprouted dual solar panels on opposite sides spanning 16.5 ft when unfolded. Mounted on the mast of the robots, science experiments charted magnetic fields, cosmic rays and dust, energetic particles, and solar plasma. In the case of Mariner 2, this suite of instruments took readings throughout the flight. Microwave and infrared radiometers were switched on as the craft neared the planet on December 14, 1962. The infrared radiometer could sense the temperature of the cloud tops – especially if they were water-bearing – while the microwave instrument was tuned to surface temperatures. One of the most important experiments was one of the simplest: observers on Earth watched

1. Possibly a dropped period or overbar when the commands were transcribed by hand.

M. Carroll, *Drifting on Alien Winds: Exploring the Skies and Weather of Other Worlds*,
DOI 10.1007/978-1-4419-6917-0_3, © Springer Science+Business Media, LLC 2011

Launch of Mariner 1 on an Atlas Agena 5 booster. The craft went off course and had to be destroyed by the Range Safety Officer less than 5 min after liftoff (NASA)

Mariner 2's radio signal as it passed behind the atmosphere as viewed from Earth. Its fading in and out – called radio occultation – revealed the structure of the atmosphere down to the surface.

The flyby made several significant discoveries. Mariner 2 determined that Venus turns in a slow, retrograde motion.[2] The planet exhibited a very weak magnetic field. Mariner also demonstrated that Venus's atmosphere is dominated by carbon dioxide, is at high pressure, and is very dry. The weather report was overcast, permanently. But the biggest surprise came in the temperature readings. Before the flyby, contemporary researchers had estimated Venus's highs to be at water's boiling point or lower. Some scientists had envisioned shorelines lapped by carbonated surf; others imagined carboniferous swamps akin to earlier – and globally warmer – epochs on Earth. Mariner 2 radioed temperatures in the neighborhood of 428°C. Venusian rocks simmered at temperatures hot enough to melt lead!

The meteorological portrait painted by Mariner 2 and the probes that followed ultimately led to our understanding of greenhouse gases. Later probes would hone our studies of atmospheric phenomena not only on Venus but as they relate to Earth.

While the Soviets set their sites on orbital and landing missions, the United States followed its Mariner 2 success with several other flybys. Mariner 5 flew within 2,480 miles of Venus on October 19, 1967. Its data

Mariner 2 carried out the first successful planetary flyby, swooping past Venus on December 14, 1962. Diagram showing the many experiments flown (left) and artist's concept (right) (NASA)

2. Retrograde planets turn east to west; most planets, including Earth, turn the opposite, or prograde, direction.

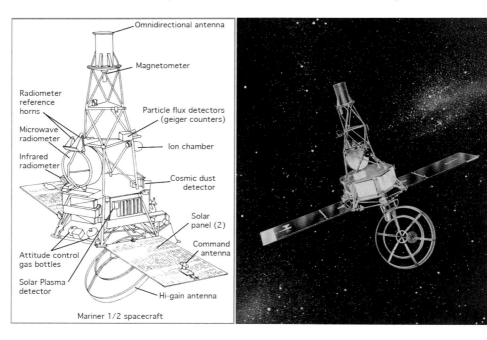

Omnidirectional antenna

Magnetometer

Radiometer reference horns

Particle flux detectors (geiger counters)

Microwave radiometer

Ion chamber

Infrared radiometer

Cosmic dust detector

Solar panel (2)

Command antenna

Attitude control gas bottles

Solar Plasma detector

Hi-gain antenna

Mariner 1/2 spacecraft

added to that of Mariner 2 and the Soviet Venera 4 atmospheric probe, which had become the first probe to enter the Venusian atmosphere just the day before.

Soviet mission planners were first to orbit the greenhouse world. On October 20, 1975, the Venera 9 orbiter became the first artificial satellite of Venus. It settled into orbit after delivering the first successful lander to another world. A few days later, the Venera 10 mother craft achieved orbit, dropping off another successful lander.

One of the first views seen from orbit around another planet, this Venera 9 image was taken on October 26, 1975 (image reconstruction by Don Mitchell)

In 1978, the Unites States joined the growing constellation of Soviet orbiters with its Pioneer Venus Radar Orbiter. This drum-shaped craft was spin-stabilized, using its rotation to keep it on track rather than a set of attitude control thrusters. Pioneer bristled with twelve instruments, primarily designed to study the upper atmosphere of Venus. Pioneer's mapping radar returned the first global overview of the lay of the land under the opaque Venusian clouds. Its resolution was about 45 miles, enough to show general outlines of continents, mountain chains, and plains. Pioneer made important discoveries about the movement and makeup of the Venus atmosphere.

Soviet engineers bettered Pioneer's radar imaging with their Venera 15 and 16 radar orbiters. Both orbiters circled the planet in elliptical orbits, allowing for mapping of the northern hemisphere from the pole down to 30° north latitude. The spacecraft could not image the clouds but did study the atmosphere through use of infrared spectrometers. The Veneras recorded chemical composition and structure of the air above the upper cloud deck, and the flow of heat escaping from below.

Orbital Venus radar systems improved with experience. Left to right: In 1978, Pioneer Venus returned a global map with resolutions of about 100 miles; Veneras 15 and 16 returned considerably better resolutions of the northern hemisphere in the 1980s; Magellan completed global mapping with better than 100 m, resolution in 1994 (NASA/JPL)

In 1989, the space shuttle Atlantis lofted the Magellan spacecraft into Earth orbit. It was the first interplanetary mission launched by a shuttle. Magellan's upper stage sent it off toward Venus. Fifteen months later, the craft

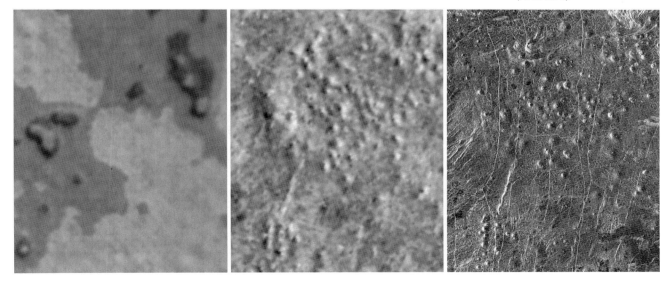

The European Space Agency's Venus Express has given the world new insights into Venusian meteorology (NASA/JPL/ESA)

dropped into orbit around our sister world. Over the next 4 years, Magellan mapped 98% of the Venusian surface at better than 100-m resolution. During four complete cycles of global mapping, it discovered that Venus has been resurfaced by volcanism within the last 500 million years. The spacecraft used aerobraking (see Box "Missions Using Aerobraking") to circularize Magellan's orbit, but there was an added benefit, says Lockheed Martin's Ben Clark. "With the Magellan mission, we did the first test of aerobraking. During that process, it allowed them to determine the atmospheric density as a function of altitude. In essence, it was an aeronomy experiment."

Magellan also studied the structure of the Venusian atmosphere by radio occultation, providing detailed snapshots of the Venusian air's density and temperature.

Venus Express, the first European mission to Venus, arrived in April 2006. From its polar orbit, the spacecraft made the first comprehensive temperature map of the southern hemisphere, uncovered evidence of past oceans, and confirmed the long-suspected presence of lightning on Venus. The orbiter also confirmed the massive "polar vortex" in the clouds over the south pole, first intimated by Mariner 10 on its way to Mercury (see Chap. 5). Venus Express has given planetary meteorologists the holy grail of orbital science: long-term study of the weather. Its mission has been extended 5 times, and it is still collecting data as of this writing.

Because Venus is so close to Earth in its size and proximity to the Sun, it is important to study. Venus and Earth began with environmental conditions far more similar than they are today. What caused Venus to become the acid-laden furnace that it is now? Climate researchers hope that Venus Express and other missions will continue to deepen our understanding of long-term climate change and greenhouse gases that are so prevalent on Venus, and on their rise on Earth. Lessons learned from other worlds repeatedly help scientists make wise decisions about the best strategies for taking care of our own environment.

SCRUTINIZING A DESERT WORLD

By November of 1964, the Soviet Union had attempted at least five Mars missions. All failed. Some failed en route to Mars. The same year that the first Soviet Venera attempted a flyby of Venus, Soviet scientists launched again. Controllers lost contact with Mars 1 en route, but other flights soon followed. Because of the way the planets move, opportunities to launch a spacecraft to Mars happen with perfect regularity. These planetary alignments, or

Missions Using Aerobraking

The concept of aerobraking has made its way into many planetary exploration proposals. The notion is that a spacecraft uses the drag of a planet's atmosphere to change its orbit, thereby requiring less fuel to slow down. It's an idea that has been kicked around for decades, and even appeared in Robert Heinlein's 1948 book *Space Cadet*, where it was used to slow a spacecraft at Venus. The Japanese probe Hiten first demonstrated the approach in 1991, using Earth's atmosphere to change its orbit. But aerobraking was not used in earnest until 2 years later, at the end of the Magellan Venus mission. Until that time, the technique was considered too risky to test on a perfectly healthy spacecraft. But Magellan's success emboldened designers to build it into their plans as a fuel-saving strategy. Below is a list of spacecraft that have used the technique.

The Magellan Venus orbiter was the first spacecraft to use aerobraking for a prolonged period. Its solar panels, seen here folded at its sides prior to deployment from the shuttle Atlantis's cargo bay, served as aerodynamic brakes in Venus's thin upper atmosphere (NASA)

Spacecraft	Aerobraking start	
Hiten (Japan)	03/19/1991–03/30/1991	First use of aerobraking
Magellan (U.S.)	05/1993	Used to circularize orbit around Venus
Mars Global Surveyor (U.S.)	09/1997	First spacecraft to rely on aerobraking as a critical part of its mission
Mars Odyssey	10/2001	Aerobraking saved 440 lb of fuel, allowing for more science payload
Mars Reconnaissance Orbiter	03/2006	445 aerobraking orbits over 5 month period

launch windows, occur every 26 months. The next opportunity would carry a spacecraft to Mars with an arrival in the summer of 1965. American space engineers met with the same fate as their Soviet counterparts on their first attempt. Mariner 3 successfully left Earth orbit, but its protective shroud stuck, keeping the craft sheltered from solar energy. Like so many Soviet probes, the spacecraft died en route.

What was it about Mars? Even with today's improved technology, fully two-thirds of all Mars missions have failed. Early engineers felt as though a dark curtain was spread between Earth and Mars, an impenetrable barrier. Why was nothing getting through? NASA personnel began to joke about the Galactic Ghoul, a beast waiting to devour any Mars-bound craft. "Conspiracy theorists were quick to propose that there really were Martians, and they didn't

Members of JPL's communications team hand color the first image of Mars to come from Mariner 4 (NASA archives)

want us spying on them. Some thought a more plausible explanation was that the shorter missions (to Venus) afforded less time for unexpected things to go wrong."

Three weeks after Mariner 3's doomed flight, the U.S. spacecraft Mariner 4 broke the spell. The craft departed for Mars on November 28, 1964, a delicate speck of chrome and sapphire tossed across the emptiness. The little probe, nearly identical to Mariner 2 at Venus, had four solar panels instead of two, and carried a camera system that took 21 full images of the Martian surface, the first images returned from deep space. Each image had a resolution of 200×200 pixels. Mariner 4 returned its treasured images at a painfully slow rate by today's standards. Each image took up to 8.3 h to transmit. As the first image came in, members of the communications team were too anxious to await the official processed image. Instead, they took strips of paper printed with numbers (Mariner converted images into bits of value numbered from zero for black to 20 for pure white) and hand-colored the image in shades of brown, approximating the true image until computer operators could assemble the real thing.

Mariner 4 revealed structure in the atmosphere above the Martian limb (NASA/JPL)

First views of Mars led scientists to believe that the planet was far more like the Moon than like Earth (NASA/JPL)

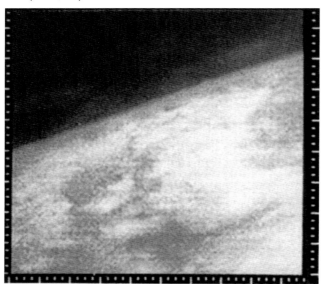

Mariner's photos took many in the scientific community by surprise. Instead of river-carved plains or Earth-like mountains, they were greeted by a barren, cratered desert world. The radio occultation experiment only added to the feeling of desolation. As Mariner passed behind Mars, its radio signals, bent by the atmosphere as they returned to Earth, showed an astonishingly thin atmosphere, said JPL space mission pioneer Charley Kohlhase. "It was a real shock when radio science occultation determined that the surface density was only 6 or 7 mbars. You know, we didn't want it to look like the Moon so much. You know, we got all these craters and we did see the edge of the southern polar cap later with Mariner 7, but it was disappointing in some ways. Everything is exciting at the time you are doing it, but you think back to the glorious notion that Syrtis Major might be some unusual biologically related feature. You remember the 'wave of darkening?' Well, all those concepts were sort of bouncing around back then."

There were more missions to come. Mariner 4 had only documented 1% of the Martian globe. But what was Mars hiding in that other 99%? It was hoped that twin spacecraft would uncover more interesting – and less Moon-like – terrain. Mariners 6 and 7 left Earth in early spring of 1969. Heavier than Mariner 4, the craft had added capability. Mariner 6 tripled the number of images returned to Earth, and Mariner 7 returned 126 photos. Mariner 6 coasted across southern temperate latitudes, while its sister was targeted to overfly the south polar cap. Both probes passed Mars at one third the distance of Mariner 4. Craters remained the theme of the encounters, although Mariner 6 spotted some chaotic terrain, and Mariner 7 took the temperature of the pole, confirming that its ice was carbon dioxide.

The Mariners seemed to have dealt a deathblow to Martian shopping malls and Lowell's canals. But scientists pointed out how warped a view an alien race would have of our own planet had they sent a probe sailing over Death Valley or the Sahara. Quick flybys of the Red Planet could only show so much. It was time to send a more capable ambassador from Earth, one that could stay for a while, chart the winds, and watch the change of seasons. The time had come for an orbiter.

In November 1971, Mariner 9 became the first artificial satellite of Mars, leaving its twin – Mariner 8 – behind in the Atlantic Ocean after a failed launch. Its mission called for a 90-day survey of Mars's meteorology, observing changes to the atmosphere and surface. Mariner 8 was to map 70% of the Martian surface. After the Mariner 8 failure, Mariner 9's mission was retooled to try to recover some goals of both spacecraft.

When Mariner 9 arrived, a planet-wide dust storm blocked its view of the surface. The only visible objects were several mysterious dark splotches in the Tharsis region. Planetary Science Institute astronomer Dr. William K. Hartmann, whose career has spanned nearly a dozen planetary missions to Venus, Mars, asteroids, and comets, was part of the imaging team. "There were four little black spots on the disk, and we were all trying to figure out what these four black spots were. I recall Carl Sagan running down the stairs to the science office from the upstairs office where the computer was at JPL.

We all had to gather at JPL because we didn't have the Internet or anything to distribute the work – the images. There was this kind of computer facility upstairs, and they actually literally took Polaroid pictures off the screen. And Carl came running down one day with this picture of one of the black spots. And it had a crater in it. That was the first moment when it began to be clear that these were volcanic mountains with a caldera sticking up above this dust layer. And from then on and through the next week, the dust at the top of the opaque dust layer was gradually settling down, down, down. And eventually we got to another interesting point where Valles Marineris appeared. Nobody had ever seen that before either; that was unknown. And it was apparent because the rest of the terrain on Mars is mostly higher so the canyon was the last thing to clear of dust. There was this period where the entire Valles Marineris complex and the outlying little closed canyons were these bright, cloudy, featureless strips of ground on Mars, because the dust was still settling in there."

As the air cleared through December, the spacecraft began mapping in earnest. Mariner 9 revealed the universality of Murphy's Law (what can go wrong will go wrong): Mariners 4, 6, and 7 just happened to have been targeted to fly over the most ancient, geologically dead regions of the planet. They missed what Mariner 9 saw: vast grand canyons as long as the continental United States, volcanoes twice as high as the highest mountains on Earth, branching river valleys, layered terrain at the poles, flood plains, and fields of sand dunes. The spacecraft survived for 17 months, returning 7,329 images. Mariner unveiled a world with rich and varied weather, both in its past and present.

Mariner 9's surface images got the most public attention, but the orbiter was equipped to be a capable weather satellite. Aside from monitoring weather patterns with its wide- and narrow-angle cameras, engineers outfitted Mariner 9 with an infrared radiometer to measure heat flowing through the atmosphere, an ultraviolet spectrometer to determine constituents in the lower atmosphere, and an infrared interferometer spectrometer to search for certain gases, to measure temperature, and to chart vertical structure.

Concurrent to the Mariner 9 mission, two Soviet orbiters made it to Mars. They arrived less than a month after the U.S. craft. Both orbiters studied the weather and returned a total of 60 images over the course of 8 months. Landings were also attempted. Two years later, Mars 5 studied the desert planet for 22 orbits, and the Mars 6 lander returned some atmospheric data in situ before contact was lost (see Chap. 4).

The Galactic Ghoul flexed its ugly talons once again; the successes of the Vikings were followed by a series of failures. The tenacious Soviet engineers designed two advanced Mars orbiters specifically targeted for the Martian moon Phobos. The clever designs utilized a laser system to vaporize samples of the moon for analysis, along with landers that could touch down on the surface. One lander was designed by Jacques Blamont, famous for his Venus balloons. Phobos 1 was lost en route, due to an errant command in a software update that commanded the attitude thrusters to turn off. Phobos 2 settled into orbit in March of 1989. The orbiter began to survey the Martian

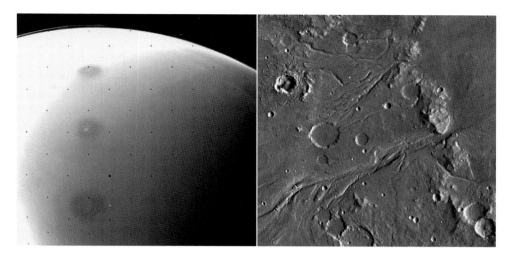

Mariner 9 was first into orbit and outlasted two Soviet orbiters. Waiting out a dust storm that unveiled only the tops of the highest volcanoes (left), the orbiter was first to show the truly varied terrain of Mars, missed by earlier Mariners (right) (NASA/JPL)

environment as it matched the orbit of Phobos. On its final approach to Phobos, just hours before landers were to be deployed, the onboard computer malfunctioned, and contact with the orbiter was lost. Two launch windows and 4 years later, NASA sent the massive Mars Observer toward Mars. With solar panels extended, the probe was 27 ft across and weighed in at over 2 tons. Mars Observer had the most advanced experiments yet to study the surface and atmosphere of the desert world.

After an uneventful flight, the craft was on final approach. Mars Observer pressurized its fuel tanks in preparation for orbital insertion and then

Soviet panoramas from the 1974 Mars 4 flyby (top), Mars 5 orbiter (center) and the 1989 Fobos 2 missions (bottom). Note the improvement in quality from 1974 to 1989. The shadow of the Martian moon Phobos is evident as dark streaks in the latter, because the Phobos orbiter followed in the same orbit as the moon. (Soviet photos reprocessed by Don Mitchell)

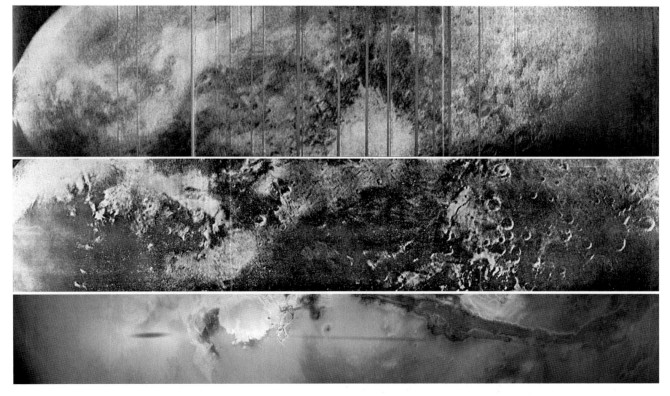

Viking orbiters – otherwise virtually identical to Mariner 9 – carried the landing craft below (NASA)

abruptly disappeared! An explosion in a fuel line was probably the cause (unless, of course, it was the ghoul).

The abyss between Earth and Mars foiled yet another mission 7 years later, when the Russian Space Agency attempted one of the most ambitious international flights to Mars. The Mars 96 mission consisted of a massive orbiter, two landers, and two penetrator probes. The network of spacecraft contained experiments and equipment built by France, Germany, the United States, and many nations of the former "Eastern Block" of European countries. The orbiter was to study climate, the abundance of various gases, ozone levels, and global weather patterns and aerosols. Each lander had a meteorology station, and the two penetrators were to study the interface between surface and atmosphere. The mother craft was an improved version of the failed Phobos spacecraft, but it never got the chance to prove itself. Due to a malfunction in an upper stage, the elaborate mission ended up as an underwater debris field off the coast of Chile.

Two U.S. successes came in the same 1996 launch window. A small lander called Pathfinder touched down on July 4, 1997 (see Chap. 4). The second success of 1997 was the Mars Global Surveyor. MGS carried some of the spare

One of the most famous Viking 1 images shows haze layers over the Argyre impact basin (NASA/JPL)

Mars Global Surveyor used its solar panels as air brakes to circularize its orbit (Corby Waste, JPL)

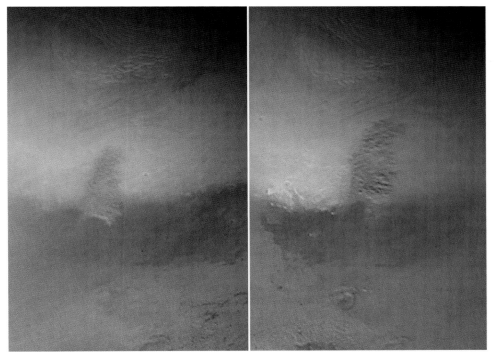

The MARCI camera aboard Mars Global Surveyor monitored weather, like this dust storm billowing over the margin of Mars's northern polar cap (NASA/JPL/MSSS)

Mars Odyssey added to our understanding of the role that water vapor and ice play in the Martian environment. Here, its THEMIS (Thermal Imaging Spectrometer) captures ice interleaved in a field of sand dunes (NASA/JPL/ASU)

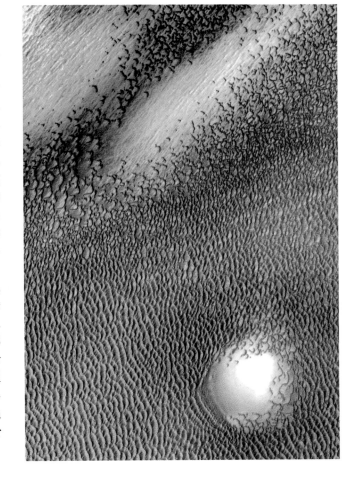

equipment from Mars Observer, though it was a smaller craft. The probe returned some 240,000 images of the highest resolution yet. Objects the size of a coffee table were visible in the images, although no coffee tables were found. Its mission lasted an impressive 9 years, unlocking many secrets about Martian weather and surface phenomena. MGS was the first Mars craft to use aerobraking (see Box "Soviet Approaches"). The MARCI (Mars Color Imager) system on the MGS acquired global views of the Red Planet and its weather patterns every day, and numerous occultations refined profiles of the Martian atmosphere.

With the increased longevity of spacecraft, and with increased data rates, handling all the incoming information became a significant issue. MGS was an incredible success, and the MGS camera team, headed by Michael Malin, knew they were going to be swamped. They recruited guest researchers who rotated through the facility in La Jolla, California, each serving for a month at a time. Malin rented a small house for the teams of three.

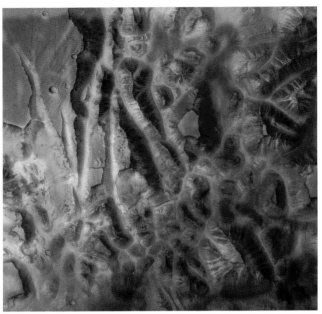

One of the spectacular images returned by ESA's Mars Express captures morning fog in the canyons of Valles Marineris (NASA/ESA)

Mars Reconnaissance Orbiter can resolve objects only inches across on the surface of Mars and is studying the weather with advanced instruments (NASA/JPL)

Next on tap was Japan's first attempt at a Mars mission, called Nozomi. The craft used gravitational swingbys of the Moon and Earth to send it into orbit around Mars. Nozomi would have orbited Mars in an ellipse, with the closest approach dipping into the outer edges of the atmosphere. But the spacecraft ran into a spate of near-disastrous problems: a stuck fuel valve dumped most of the gas intended for use at Mars; solar flares damaged the communications and power subsystems; an electrical short – probably due to the solar activity – caused a failure in some heaters, allowing fuel to freeze. After thawing the fuel en route to Mars, the craft became disoriented at the time it was supposed to fire thrusters for orbital insertion. Despite heroic attempts to save the mission, Japanese engineers ultimately had to abandon their first Mars mission in December of 2003.

Japan's disappointment was not the last at Mars. The U.S. Mars Climate Orbiter departed Earth under NASA's new theme: faster, better, cheaper. Because of shortcuts in personnel, and miscommunication between JPL and the prime contractor, Lockheed/Martin, flight navigators sent the capable orbiter too close to Mars for orbital insertion, and the craft burned up in the atmosphere it was supposed to study. JPL and Lockheed/Martin each used different units of measurement (metric vs. English), leading to the disaster. The doomed Mars Polar Lander (Chap. 4) was lost during the same launch opportunity.

Three new orbiters are currently scrutinizing the Martian environment in great depth. First to arrive was Mars Odyssey, which has been mapping the planet since 2001, discovering caves on the flanks of volcanoes, CO_2

geysers in the Martian polar caps, and water-related material in rocks. The European Mars Express orbiter arrived in December of 2003. It has returned thousands of spectacular 3D images of the surface, moons, and atmosphere. The spacecraft also discovered concentrations of methane in several locales. Methane can only remain stable in the Martian environment for about three centuries before it dissipates and combines with other gases. Something must be replenishing it. Scientists have put forth two theories: either the methane is evidence of recent volcanism, or it is related to life. Researchers are awaiting more detailed results. The Mars Reconnaissance Orbiter is charged with studying the history of water on Mars.

Each orbiter is keyed to studying a different aspect of Mars, but the Mars Reconnaissance Orbiter is the most versatile craft yet to orbit the Red Planet. It carries the most powerful camera system ever flown above the Red Planet, says MRO imaging scientist Steve Lee. Lee is based at Denver's Museum of Nature and Science, and also works at Boulder's Laboratory for Atmospheric and Space Research. "It's the coordinated observations you can do with MRO that makes it so powerful. It's got a suite of instruments that cover from the MARCI scale – roughly three quarters of a kilometer per pixel globally, so you get a global picture of what's happening – all the way down to the HiRISE camera, which gives you 20 cm per pixel. These things are all sort of boresighted, if you will, so that when you're looking at something with HiRISE, you're also seeing sort of a regional context with MARCI. And then you've got the context imager, which is 6 m per pixel, so Hirise is nested within these ever increasing spatial resolution data sets. And then in addition to that, you've got instruments like CRISM, which is the imaging infrared spectro-meter, and so you can come up with spectral information on this as well." The orbiter will also serve as a communications relay satellite for future missions such as the upcoming Mars Science Laboratory rover mission.

Soviet Approaches

Viktor Kerzhanovich has had the opportunity to work with Soviet, Russian, and American planetary programs during the period considered by many historians to be the "golden age" of planetary exploration. From 1963 to 1981, he worked at the Russian Scientific Research Institute for Space Instrument Engineering. In 1981, Kerzhanovich transferred to the Institute of Space Research (IKI) of Russian Academy of Sciences. He left in 1994, and today works at NASA's Jet Propulsion Laboratory in Pasadena. During the early days of explo-ration, workers on both sides of the Atlantic had to deal with boundaries set not only by the Solar System but also by their respective governments. Here, Kerzhanovich shares what those days were like.

"[In the early days of planetary exploration] it was so exciting what was being accomplished. You can think about it – you can dream about it – but you have to be more or less realistic about your capabilities. At those times, research was practically unlimited, but the technical level was limited at all entities of ours. That's why we had problems with Mars and there was no Soviet probe sent to the outer planets. For the Moon, yes, for short periods we could do it because all the reliability of our avionics and many other things were far from reliable to get several years of flight time. But it was hard in those times, because in those times we could build everything made of only Soviet-built components. You cannot buy anything and so you cannot compete against the world. But a lot of exciting things were being made in those times. You remember there was [the *Luna* series] lunar landing and lunar sample return. And we had the landings on Venus surface. There was sampling from Venus surface. There were attempts on Mars, and you know the first rover was landed on Mars, but the landing was very hard. This was in 1971. It was about the size of the Sojourner rover. So it was quite exciting."

GLIDING THROUGH GIANTS

The year 1979 was fast approaching, and with it, a unique cosmic arrangement. It happens rarely, just once every 179 years. And it was an event that NASA could not let pass without at least an attempt.

The planets each travel at their own speed in their concentric track race around the Sun. The further from the Sun, the more slowly they must move to stay in solar orbit. While Jupiter takes just under 12 Earth-years to make its annual circuit, Neptune's year lasts nearly 165 years. Differences in orbital speeds contribute to varying alignments of the planets, as one overtakes, then lags behind, another. A spacecraft traveling through the outer Solar System can take advantage of these changing arrangements using the gravity of one planet to get to another. This technique is called gravity assist. In a cosmic game of billiards, the craft can be targeted precisely enough to use one planet's gravity as a bank shot to another. And in 1979, all the outer planets would be aligned in such a way that one spacecraft could, theoretically, visit all four worlds. Jupiter's gravity would bend the craft's trajectory so that it could continue on to Saturn, then to Uranus, and finally past Neptune. This decade-long journey carried the name of "The Grand Tour." But would it work?

Carolyn Porco, Imaging Team Leader of the NASA/ESA Cassini Saturn Mission, said: "No matter how you measure it, whether you count the number of bodies, whether you measure the volume that is taken up by the orbits, whether you talk about the amount of mass in the outer Solar System, no matter how you cut it, most of the Solar System exists beyond the orbit of the asteroids." If the outer planets were to be explored separately, mission flight times were so long that it might take half a century to reconnoiter the entire Solar System. The Grand Tour would solve that problem with one launch, one spacecraft, and one journey to all four planets. But the journey was a treacherous one, with unknown variables. In the early 1970s, Jupiter seemed like an impossibly distant target. In the 1960s, spacecraft had crossed the 100 million mile abyss to reach Venus. By the middle of the decade, they had ventured twice that far to get to Mars. But no spacecraft had traveled beyond the orbit of the Red Planet. Out there lay the giants in the land, unseen hazards waiting to destroy a delicate spacecraft.

The first danger lay in a string of deadly pearls known as the Asteroid Belt. The asteroids orbit throughout the Solar System, but most circle the Sun in a donut between the orbits of Mars and Jupiter. Nothing in the Asteroid Belt is big enough to be considered a planet. The largest object, Ceres, measures just 930 km across, and yet it weighs in at a quarter of the mass of all asteroids combined. Asteroids range from stony to metallic in composition. By the year 2000, over 8,000 had been charted. Researchers estimate that more than 50,000 asteroids larger than 1 km are orbiting the Sun within the Asteroid Belt, many unseen. Any spacecraft on its way to the outer Solar System would have to cross this gauntlet.

The second lethal roadblock to a Grand Tour lay close to Jupiter itself. Jupiter's magnetosphere is second only to the Sun's. Many planets have a magnetic "bubble" surrounding them. Earth has its Van Allen radiation belts and other shells of energy, and it is not alone. Planets that rotate fairly swiftly and contain an electrically conducting fluid core where convection is occurring seem to generate these energetic fields. The strongest by far is Jupiter's. Close to the planet, radiation is fierce enough to destroy electronics and scramble computer brains. Unfortunately, in order to use Jupiter's gravity to deflect a spacecraft on to Saturn for its Grand Tour, the craft had to be targeted well within these deadly fields. Could a probe be built to survive such a storm of radiation?

The outer planets presented yet another obstacle – power. Halfway to a destination, sunlight drops off by 4 times. Solar energy becomes feeble beyond Mars. Both NASA and the Soviets began experimenting with nuclear power. Outer planet spacecraft do not carry nuclear reactors, but rather sophisticated plutonium power generators that convert heat to electricity. The most common nuclear power on U.S. spacecraft is called a radioisotope thermoelectric generator, or RTG. The use of nuclear power for space exploration revolutionized outer planets mission scenarios, says JPL's Charles Kohlhase, who designed and managed several major outer planet projects. "It made it possible to go to Jupiter and beyond. We could fly missions out to Mars without having unmanageable solar arrays. But when you went to Jupiter, although you could conceivably use solar arrays, they would need to be huge. A mission like Mariner 4 or 6 or 7 had to use less than the Voyagers used and the Voyagers got by with under 300 W. If a Mariner was using 100 W – an early Mariner – its solar arrays were something like one by three meters. If you then go out to Jupiter, which is 5 AU (astronomical units) instead of 2, you've got to square the energy you get from that distance, so you need about 6 times more [solar panel area]. Now, if the power you need isn't too much, you're up to 30 m^2, but if you've got to run some high power stuff, pretty soon you get to the size of a tennis court. And now you start to have reliability issues just unfolding all that and managing it. Once you begin to get to Jupiter and beyond, it's just kind of a pain to manage the size of the solar array. You've got to go to something efficient and compact. And along comes plutonium dioxide, or RTGs, which are radioisotope thermoelectric generators. They're small and compact, and they convert the heat of radioactivity into electricity through thermal couples. As long as you can get approval to launch them, and not put anybody at risk, they last a long time. Their half-life is about 89 years, so it made a lot of sense to use them."

Using nuclear power of any kind is controversial. NASA had to do many varied tests to prove that in a launch accident the plutonium in the RTGs would remain intact, Kohlhase says. "The term plutonium scares everybody to death, but in the RTGs it's not in its metallic form. Plutonium dioxide is sort of like a ceramic dinner plate. It's like spheres. And if they break, it doesn't really do any harm. You have to sort of pulverize them into fine dust and spread it all around before it's a [health or environmental] issue. Usually

Pioneer 11 encountered Saturn on September 1, 1979 (J. Higgins, Wikipedia commons)

most accidents don't do that. They might break up but they probably aren't going to be distributed widely in the atmosphere." Kohlhase points out that solar array missions to Jupiter require huge, ungainly solar panels. A spacecraft at Saturn – which is twice as far away – requires 4 times as much area. "At that point you just say let's go nuclear."

Scientists had their first outer planets flying practice with two probes in the early 1970s. Pioneers 10 and 11 both flew by Jupiter, and Pioneer 11 also encountered Saturn. Pioneer 10 arrived at the Jovian shores in December of 1973. Pioneer 11 passed closest to the planet December 2, 1974, and made it to Saturn for a close encounter September 1, 1979. Carolyn Porco says, "Pioneer was basically a mission to see if we could even survive passage through the Asteroid Belt and skimming the Jupiter system, going through its magnetic environment to see if we could live. The camera – which was rudimentary – was an afterthought."

Like all outer planet spacecraft, the Pioneers were dominated by a giant dish antenna to communicate with Earth over vast distances. Pioneer 10 and 11 were spin stabilized. Instead of locking on to several stars and keeping the spacecraft oriented by the use of tiny attitude control jets, the craft were set

Pioneers used the spinning movement of the spacecraft to scan objects. The strips of images radioed back to Earth had to be corrected for this movement (NASA/JPL)

spinning like a top. A spinning object tends to remain pointing in one direction. Spin-stabilized spacecraft require less fuel, but they have the drawback of constantly rotating. Some experiments are perfectly well suited for this constant panning across the sky. But instruments such as imaging systems prefer to be pointed in one direction. The imaging system on Pioneer had to use the motion of the spinning spacecraft to scan lines of images. This made for less resolution than a stable platform could provide. Still, for an afterthought, the imaging system did well. Pioneer revealed unseen structures at the poles and better detail than was possible at ground-based observatories at the time.

Pioneer also showed, for the first time, how evenly heat was distributed across the planet. Pioneer's infrared radiometer was the instrument that first showed the flatness of temperature between equator and pole. That global internal heat flow creates upwellings that drive many of the weather features on the king of worlds. Pioneer 11 returned spectacular images of Saturn's rings, discovering an unknown one in the process. The craft also discovered two new moons and took the temperature of Titan, finding it as cold as the lowest estimates. Lunar and Planetary Laboratory scientist Peter Smith was part of the Pioneer team. "I was in the control room up at Ames doing the flybys. It was really fun. Pioneer 11 arrived in September of 1979. For a few weeks beforehand, we were in the control room where we would send commands up and receive data back from the spacecraft. There was 1½ h of one-way light time, so it was 3 h from when the commands went up to when the data came down. It was really very interesting. We had these very primitive display methods, nothing like we have today. It was almost like oscilloscopes displaying the data."

The Pioneers were the first spacecraft to be launched fast enough to escape the gravity of the Solar System. Each carries a plaque showing the spacecraft's point of departure on a tiny planet next to an average star tucked away in a corner of the Milky Way Galaxy.

While the Pioneers were still under construction, NASA lobbied for their sophisticated "Grand Tour" multiprobe spacecraft, but Congress balked at the price tag. Instead, NASA redesigned its reliable and highly successful Mariner spacecraft, outfitting it with a huge antenna and twin radioisotope thermoelectric generators. The new craft cost a third of what the Grand Tour would have. Project planners christened it Voyager.

Technicians built two Voyager spacecraft. Voyager 1 was launched on a trajectory that would carry it past Jupiter and on to Saturn, with an emphasis on Saturn's atmosphere-clad moon Titan. If the

Twin Voyagers carried out the first detailed reconnaissance of the outer planets. Voyager's main antenna is 12 ft across (NASA/JPL)

spacecraft remained healthy, Voyager 2 would be targeted to do a scaled-down version of the Grand Tour, passing Jupiter and Saturn before traveling on to Uranus and Neptune over the course of a decade.

In charge of atmospheric studies on the bold venture was California Institute of Technology's Andrew Ingersoll. The science of planetary atmospheres was still in its early stages when the Pioneer missions to Jupiter and Saturn came along. The reconnaissance of the outer planets was a stunning experience of discovery. In fact, nothing prepared them for the spectacle awaiting the Pioneers and Voyagers, Ingersoll says. "We knew that these planets were gas giants at that point, and we knew that the features we saw were clouds. We knew that they were remarkably steady and long-lived. After all, the Red Spot had been observed for hundreds of years. It wasn't quite so clear how stable the jet streams were, but by Earth standards, everything was really pretty steady and long-lived. For a meteorological puzzle, it was a pretty good one because all our intuition said that here, flows are turbulence and they break up, based on Earth's analogy. Then – to some extent Pioneer, but much more Voyager – deepened the mystery by getting closer so we were able to see the small scale motions. They were quite turbulent, and it was even a bigger paradox to see how large structures like the Red Spot and the large ovals, which also have a long documented pedigree, and the jet streams could just stay there without breaking up. It deepened the mystery."

Following its encounters at Jupiter and Saturn, Voyager 2 ventured on to Uranus. The encounter was the first at the green giant, and the most distant of any up to that time. The path followed by the spacecraft was quite different from any other because of Uranus' odd tilt. The planet is tipped over by 98°, essentially spinning on its side. Voyager approached from Uranus's south pole. From Voyager's standpoint, the rings and moon orbits around the planet were arranged like a giant bullseye.

After its harrowing journey (see Box "The Rocky Road to Neptune"), the spacecraft performed beautifully. Engineers and scientists were still coaxing discoveries from the encounter 4 years later as Voyager approached the most distant gas giant, Neptune. Traveling at 12 times the speed of a rifle bullet, Voyager sped through the system, discovering new moons, studying Neptune's mysterious ring arcs, and mapping storm systems on Neptune. An added payoff came in the form of an encounter with Triton, Neptune's large, atmosphere-engulfed moon (see Chap. 9). Voyager had completed four complex planetary encounters over the course of 12 years, discovering dozens of moons and revealing magnetic fields and weather systems never before known.

The Pioneer and Voyager missions had proved that gravity-assisted flight plans worked. Soon, most planetary missions were taking advantage of the cosmic pinball game that is our Solar System. One of the first flight engineers to work on such trajectories was JPL's Louis Friedman, who later founded the Planetary Society with Carl Sagan and Bruce Murray. "That was a really active time for celestial mechanics folks for coming up with these fantastic orbits that, at first, were just multiple planet flybys, like Mariner Venus-Mercury. Then there were grand tour flybys with multiplanets. Then there were Galileo-type missions where you could change your orbit in doing different encounters with it, and change the orbital plane. And then the whole discovery of all the Vega trajectories (Venus-Earth gravity assist, which used Venus's gravity to slingshot a craft to the outer Solar System)."

The added boost from gravity-assisted flight paths allows for much heavier spacecraft such as Galileo/Jupiter and Cassini/Huygens/Saturn. Every 10 years, NASA embarks upon what is known as a "Flagship Mission," a multi-billion-dollar, international interplanetary project. Jupiter was the target of the flagship mission of the 1990s, and the Galileo Orbiter/atmospheric probe was the spacecraft.[3] Like the Voyagers before it, Galileo changed the very foundations of our understanding of the Solar System. The orbiter was a hybrid of previous spacecraft. Galileo combined the best of two worlds, with a spin-stabilized body to save fuel over its long mission, and a despun section that carried scientific instruments.

Plans were being drawn for the Galileo orbiter and probe in the early 1970s. NASA Ames began formal designs in 1975, with a target launch aboard a space shuttle in 1982 using a powerful Centaur upper stage fueled with cryogenic liquid propellants. The flight would have also encountered Mars on its way, gaining a gravity assist from the desert world. But disaster visited NASA's space program in 1986, when the space shuttle Challenger exploded on ascent, killing all the crew aboard. As experts studied all aspects of the shuttle to make it safer, the Centaur was deemed too dangerous to carry in the cargo bay. Instead, Galileo would leave a shuttle atop an inertial upper stage, a solid rocket that was far more stable and safe, but far less powerful.

Launch finally came on October 18, 1989. Galileo had a price to pay for using a weaker upper stage – its new course required help from the gravity of several planetary flybys. Designed to operate in the frigid outer Solar System, Galileo now had to also endure one flyby of Venus and two of Earth before its looping orbit took it toward Jupiter. Designers added a Sun shield to protect the craft, but the long trip and higher temperatures at the start resulted in a major failure. Galileo's umbrella-like main antenna jammed. Despite efforts to alternately heat and cool it, and even to jostle it open, the antenna wouldn't budge. Instead of a mission using a 16-foot diameter high gain antenna, Galileo now had to rely on a tiny low-gain antenna the size of a pie pan. The data rates for transmission to Earth were drastically reduced, and some experiments had to be curtailed. Thanks to the creative and heroic efforts of flight engineers and scientists, the mission proceeded. The computer was taught to compress data in new ways, and various encounters were redesigned to be more streamlined.

Engineers encrusted Galileo with an impressive battery of instruments. A solid-state imaging system returned detailed images in parts of the spectrum near the visible realm. A Near-Infrared Mapping Spectrometer determined the composition of the moons and Jupiter's atmosphere. Galileo's Photopolarimeter-Radiometer could see not only heat reflected off surfaces

3. Galileo mission cost $1.35 billion (US) over twenty years of development and flight. The Cassini/Huygens project came in at just under $3.3 billion (US).

Galileo being assembled in the clean room at JPL. The despun section is in green (NASA/JPL)

from sunlight but also heat coming from the interior of the planet and its moons. The instrument also charted methane and ammonia in the clouds. An Ultraviolet Spectrometer studied the aurorae and molecules in the upper atmosphere. Flight engineers also observed radio signals at each occultation. A host of other instruments observed radiation and dust in the Jovian environment.

The Galileo orbiter logged many discoveries, including the close encounters of several asteroids; evidence of subsurface oceans on Jupiter's moons Europa, Ganymede, and Callisto; new and continuing volcanic activity on Io; massive thunderstorms and "hot spots" on Jupiter; and winds of over 400 miles per hour. (For a full weather report, see Chap. 7.)

The first decade of the third millennium saw a flagship mission to Saturn's vast system of rings and moons. Arriving in 2004, the 6-metric-ton Cassini orbiter settled into orbit around Saturn after a 96-min burn of its main engine. Just 150 days earlier, Cassini had gently ejected the Huygens probe on a slightly different path that would intersect Titan. The separation enabled Cassini to relay data from the probe to Earth in real time as the Huygens descended through the dense atmosphere of Titan (see Chap. 4).

Despite a fouled main antenna, the Galileo carried out a complex multi-year mission in orbit around Jupiter (© Michael Carroll)

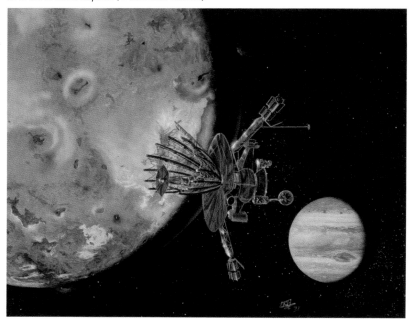

In addition to surveying magnetic fields, moon surfaces, and rings, Cassini is equipped to be a formidable weather station. Its composite infrared spectrometer (CIRS) measures the heat and composition of surfaces and clouds. The Ultraviolet Imaging Spectrograph (UVIS) senses compounds and gases that other instruments cannot. It maps hydrogen, oxygen, methane, water, and other elements in the atmospheres of Titan and Saturn. The visual and infrared mapping spectrometer (VIMS) and the Imaging Science Subsystem (ISS) contribute by detecting gases and compounds in other parts of the spectrum.

Cassini is the most complex spacecraft ever flown. Although the craft scrutinizes many aspects of the Saturn system, its primary goals are twofold: to study Saturn and to study Saturn's planet-sized moon Titan. The majority of Cassini's circuits around Saturn bring it close to Titan. As it flies by, the VIMS and ISS can image the surface through the haze, while its entire battery of experiments studies atmospheric haze layers, clouds, and surface features. Cassini also beams radar through the atmosphere to map the surface in great detail, much as the Magellan did at Venus in the 1990s. Finally, when Cassini passes behind the moon as viewed from Earth, researchers monitor the changes in its signal to chart the structure of the atmosphere. This is also done with Saturn itself, and the same technique is used to search for rarified atmospheres around other moons.

NASA's Cassini orbiter, as large as a school bus, delivered the European Space Agency's Huygens probe to the atmosphere and surface of Saturn's moon Titan (NASA/ JPL/Michael Carroll)

Flybys of the planets were the first logical step in planetary exploration. All a spacecraft had to do was to survive long enough to get fairly close and transmit its findings home. Once this technique was mastered, the next step was to orbit a planet. Orbiters enable long-term study of a planetary environment, from its magnetic fields to its weather, seasons, and climate. But to experience a planet's weather firsthand, probes had to venture down into the atmosphere and drift on alien winds.

Cold War Secrets

Often, the search for knowledge crosses international borders and trumps politics. This was certainly the case when Pioneer made humankind's first reconnaissance of Saturn in 1979. Planetary atmosphere veteran scientist Andy Ingersoll had an experiment aboard.

"With Pioneer 11 at Saturn, there was a low-level secret agreement between Soviet and American engineers that the Soviets would maintain radio silence while we were gathering the signals from Saturn, which were pretty weak. Neither side wanted to divulge this agreement because their bosses would have gotten angry. But the engineers were very cooperative; everyone was very friendly at the engineering level. The U.S. side had told the Soviet side that they could turn on their Earth-based communication and spy satellites at a certain hour. They had forgotten that my experiment was still gathering data from Titan. It was early in the morning at Ames Research Center, and I was with one guy and my programmer in this vast room with television monitors. All of a sudden, all these monitors started flashing in big block letters, and it looked very bad. They technician who ran the show was in contact with Canberra Deep Space Network (in Australia) said, "It's radio frequency interference." None of us in the room knew where it was coming from, and none of us knew about this secret agreement. The next day, they had to explain what had happened to the Titan data at a press conference. They just said, "We just had RFI," which didn't explain anything. But the *National Enquirer* said, "We know what happened. It was the residents of Titan jamming the signals so that the infrared radiometer would not detect the heat of our cities." That was their theory. It's a great story, and the fact that both sides were deathly afraid that their bosses would find out about cooperation, that's a nice part of the story."

The Rocky Road to Neptune

The twin Voyager spacecraft were the first to be designed with a lifetime of over a decade. Voyager's complex mission, coupled with the immense distances across which the craft would need to communicate, required an unprecedented amount of autonomy. At Neptune, Voyager 2's radio signals would take 4 h to reach Earth. If an emergency occurred, it would take another 4 h for flight controllers to get a command to the distant craft.

The Voyagers were programmed to periodically take inventory of their own health, and could switch to backup systems as needed. One such contingency plan involved the radio. Each time the craft received a transmission from Earth, an internal clock reset. If another command was not received within a certain amount of time, the craft automatically switched to a backup receiver and called home. In fact, Voyager 2 needed its backup receiver when both fuses blew in the primary. But the backup receiver had its own problems. A critical component of the communications system was called the tracking loop capacitor (TLC). This device sensed the frequency of an incoming signal and tuned the receiver to that frequency. When the TLC failed, Voyager was essentially deaf. Flight engineers had to figure out the correct frequency each time they sent a command, but this was not easy. The frequency of a traveling spacecraft continually changes. The Voyager team had to calculate the motion of the tracking station on the turning Earth, the motion of Earth as it related to the Moon, the motion of Earth around the Sun, and the velocity of the spacecraft relative to Earth and the Sun. They successfully did these calculations each time that a command was sent over the course of Voyager's 12-year flight.

Other problems plagued the plucky craft during its odyssey. The scan platform, a system for pointing the camera, partially jammed at Saturn. Encounters at Uranus and Neptune had to turn the spacecraft itself to target images. An infrared instrument was temporarily blinded by fog from material that bonded its mirrors in place. Engineers had to figure out how to heat the mirrors to defog them before each encounter. Filter wheels, which enable instruments to see at different wavelengths, also jammed as lubricants drifted away in the zero-g environment. Despite these and other challenges, flight engineers at JPL were able to coax the aging craft through spectacular encounters with all the outer planets, a Grand Tour indeed!

A violent downdraft catches the Soviet VEGA 1 balloon. The Soviet/French-built probe was the first balloon to sail on the alien winds of Venus (© Michael Carroll)

Chapter 4

Studying the Weather from the Inside

THE TERRESTRIAL PLANETS

Gazing across the vast distances of the cosmos, two planets seemed likely targets for early atmospheric explorers, Mars and Venus. Both were fairly accessible, being the closest planets around. Venus was an easier target to get to. The planet's path takes it around the Sun at an average distance of 67 million miles, compared to Earth's 93 million miles. At times, the planets are a scant 23.7 million miles apart. An added attraction of Venus is that it is "downhill." Things in the Solar System tend to fall toward the Sun. A spacecraft is no different. Since Venus is closer to the Sun than Earth is, the Sun's gravity helps speed a Venus-bound spacecraft, while a Mars-bound one is fighting against the Sun's gravity, going outward. All these factors meant that a trip to Venus would take roughly 4 months, while a trip to Mars could take as long as eight. In the early days of space exploration, a few extra months often meant the difference between success and failure.

Little was known about conditions beneath the clouds of Venus when the Soviet Union and the United States attempted the first missions. Soviet designers came up with a series of spacecraft dubbed Venera (meaning "Venus"). The probes were designed to study the atmosphere on the way to a landing, but no one guessed how much stress the probes would need to withstand to do so.

After several failures (see Appendix 1, "Venus Missions"), Venera 4 successfully traversed the airless abyss separating Earth and Venus. It was a daring event, a Promethean task the engineers and scientists tried to undertake, defying the vast chasm set down between the planets.

Venera 4 survived the initial fires of entry, dropped its protective shell, and deployed its parachute. The probe was designed to withstand 10 atm of pressure (10 times the pressure of Earth at sea level). Venus crushed the unsuspecting craft 16.2 miles above her surface, at a pressure of approximately 15 atm.

Undaunted, Soviet engineers at the Lavochkin Institute – which led the Soviet planetary effort – crafted reinforced versions. Venera 5 and 6 each made it into the atmosphere, surviving down to about 12 miles before succumbing to pressures above 25 atm. Soviet designers began to wonder how bad the pressure could get on our sister world. Scientists combed through the data and helped engineers to decide on the next design, one that could withstand up to 180 atm. Each craft was nested within strong titanium shells. The added strength made the probes so heavy that equipment was removed from the upper stage's telemetry gear.

Venera 7 became the first probe to make it to the surface of another world. During its descent, the craft radioed temperature profiles (other experiments failed) until it reached a few yards from the surface, when contact abruptly ceased. Later studies revealed that the craft continued transmitting a weak signal for 23 min. Analysts believe the parachute catastrophically failed in the final stages of descent, and that Venera 7 landed on its side. The probe proved

M. Carroll, *Drifting on Alien Winds: Exploring the Skies and Weather of Other Worlds*,
DOI 10.1007/978-1-4419-6917-0_4, © Springer Science+Business Media, LLC 2011

Far into the future, the Venera 9 lander will remain undisturbed beneath the brooding acid hazes of Venus. The craft landed on a dramatic, rugged, 30° slope of volcanic rock (© Michael Carroll)

that a spacecraft could survive the rigors of Venus' hellish environment. Venera 7 sent temperature readings of over 860°F. Though Venera's pressure sensors could not function due to an electrical problem, pressure can be computed as a function of temperature. Scientists estimated pressures equivalent to those half a mile below the surface of Earth's oceans (93 atm, or 93 times as dense as Earth's air at sea level).

Venera 8 followed in 1972, carrying sophisticated equipment to study Venusian weather. The craft was equipped with a photometer to measure light levels in and between clouds, a mass spectrometer to analyze gas types, temperature and pressure sensors, and a gamma ray spectrometer to measure composition of the surface – if it made it that far. It did. The craft survived for over 50 min on the surface.

Veneras 9 and 10 – delivered by mother ships that went into orbit – were identical probes designed to truly operate on the surface of Venus. They carried a host of instruments, including thermometers, barometers, mass spectrometers, gamma ray spectrometers, soil density probes, and cameras in hopes of giving us our first glimpse of the surface of another world. The craft also carried nephelometers, essentially lights that detected fog density by looking at the brightness of their own beam in the air.

Venera 9 was the first of the twins to travel the winds of Venus. As it passed the 40 mile altitude above the Venusian deserts, three parachutes opened. The descent through the clouds enabled measurements of wind, temperatures, water vapor, and pressure. Venera confirmed the hurricane-force winds at high altitudes. For 20 min, Venera passed through Venus's three cloud decks. At about thirty miles, Venera jettisoned her parachutes, descending into thicker air. The craft now relied only on a metal disk, an aerodynamic "brake," to slow it. As it neared the surface, the atmosphere became as dense as water, and winds died down to a gentle breeze of 3 m/s. Venera 9 became the first probe to operate on the surface of another world, setting down in the highlands of Beta Regio. For the next 53 min, Venera returned haunting images of a rocky slope simmering at nearly 900°F under an overcast sky. Venera's panorama covered a nearly 180° field of view, and it was able to sample the soil.

These beautiful views from the Soviet Venera 13 (left) and Venera 14 (right) landers were reconstructed from the original Soviet 1982 files (bottom from Venera 13) by Don Mitchell. Note the clarity of atmosphere and the high light levels, both a surprise to planetary meteorologists (© Don Mitchell, http://www.mentallandscape.com)

From 1975 to 1981, a veritable Venera armada gathered to investigate our sister world. Carried by successful orbiters, Veneras 10, 11, 12, 13, and 14 all made it to the surface with varying degrees of success. Veneras 13 and 14 transmitted stunning color panoramas of their surroundings, as well as data on soil and weather.

In the midst of the Venera parade, American scientists dispatched their Pioneer Venus mission. The Pioneer Venus orbiter arrived shortly before a quartet of atmospheric probes made their way through the sulfuric acid-laden air. The Pioneer Venus probes were targeted to enter in four different areas and were named according to their destination. Three small probes were labeled North Probe, Day Probe, and Night Probe. Each was about 3 ft in diameter, had sensors for temperature, pressure, and acceleration, along with a nephelometer to sample the air and an experiment to monitor heat flow through the atmosphere.

A fourth probe, called the Large Probe, carried seven instruments in a 4-foot diameter pressure vessel. Unlike the free-falling small probes, the large probe descended by parachute. All probes were used to track the winds. None were designed to survive a landing, although the Day Probe continued transmitting from the surface for over an hour.

The four probes in the Pioneer Venus Multiprobe mission traveled the interplanetary void atop a bus that acted as an upper atmosphere probe as it burned up, after jettisoning the atmospheric probes toward four destinations on the planet (NASA/Ames)

The four probes were delivered to Venus by a bus that powered them through the long journey from Earth to Venus. The bus itself contributed to studies of the upper atmosphere. It was not designed to survive entry, but instead relayed real-time measurements using mass spectrometers to sense composition of the upper fringes of the Venusian air. This region of the atmosphere had never been studied directly, as all previous probes waited until after entry to deploy their experiments. As planned, the bus burned up at an altitude of about 65 miles.

Soviet scientists had enjoyed tremendous success at Venus, but they were not content to have their Venus probes remain locked to the ground. When the opportunity arose, Soviet engineers would try something different, something remarkable, and something that harkened back to eighteenth-century French paper-makers.

That time came in 1985, when two advanced probes arrived at Venus with a second destination in mind – the comet Halley. VEGA, which stands for the Russian combination Venus/Halley, consisted of two flyby spacecraft and two landers. After dropping off the landers, the flyby spacecraft continued on to encounter Halley's Comet. The landers themselves were nearly identical to the Venera landers with one striking exception – each carried an "aerostat" balloon probe designed by NPO Lavochkin and French aerospace engineer Jacques Blamont of France's Center for Space Research (CNES). As the landers descended, they released their helium-filled balloons some 33 miles above the ground. The altitude was important, says Blamont. "Venus ballooning is very easy above 50 km [31 miles] altitude. It is cooler up there. At lower levels, it is still possible, but has not been done." At an altitude of 39 miles, each of Blamont's aerostats pulled away from the lander on its own parachute. A second parachute pulled the balloon out 27 miles up. Air tanks slowly filled the Teflon envelope. The balloon inflated 100 s later, bobbing back up to a cruising altitude of between 30 and 32 miles. The pressure at that height was 532 mbar, half what it is at Earth's surface. Upon inflation, the parachutes and air pressurization tanks dropped away.

Each Teflon balloon spanned 11½ ft in diameter and carried a battery-powered instrument gondola 42 ft below. The gondola was 4 ft tall and carried experiments to observe pressure and light levels. An arm extended to measure temperature. The arm also held an anemometer to track windspeeds. Below these

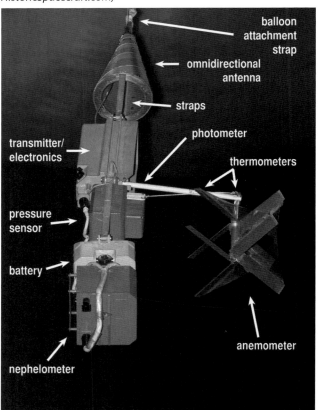

The meteorology gondolas hung below VEGA's 11½-foot Teflon balloons and survived for 2 days in the harsh Venusian environment (© Richard Kruse, HistoricSpacecraft.com)

balloon attachment strap

omnidirectional antenna

straps

photometer

thermometers

transmitter/ electronics

pressure sensor

battery

anemometer

nephelometer

was a nephelometer, essentially a light source that observed how much light was reflected back by mist.

Twenty ground stations tracked the movement of the two probes over the course of their lifetimes. The probes each lasted for more than 46 h, floating in the middle cloud deck of Venus. This region is the most active atmospheric layer. The balloons coasted west at an average speed of 154 miles per hour, sometimes buffeted by strong vertical drafts. Temperatures hovered at 98°F, comfortable considering the hellish conditions far below. Each balloon traveled approximately 6,890 miles before ground stations lost contact.

The balloons experienced differing temperatures. The discrepancy in the two VEGA balloon temperatures is somewhat mystifying. Some scientists feel the difference is due to irregularities on the surface of the planet below. But veteran Venus researcher Kerzhanovich is skeptical of that scenario. "The difference was about 15°. It happens on Earth, and nobody expected it to happen on Venus. I do not think it is related to topography, but it is related to where the air masses came from. I would say it is sort of a mystery, but unlikely that the surface is the cause. The surface is too far. It's about 55 km below. It's unclear what happened." The details of this and other Venusian mysteries may be solved by future explorers now on the drawing board (see Chap. 10).

In the Heat of the Moment

When the forecast calls for daily highs hot enough to melt lead, one needs very special paint to color things. Each Venera lander carried a strip of colors so that camera images could be calibrated. But the complexities of space missions breed miscommunication on such things. Scientists, engineers, navigators and technicians all have different focuses, and those divergent goals are sometimes translated into conflicting solutions to problems, solutions that can result in everything from minor nuisances to mission losses. This phenomenon befell the Venera landers of the Cold War Era, according to planetary scientist Bill Hartmann, who discussed the incident with a Soviet scientist over a decade after the occurrence played out.

"In those days, the Russian system was divided; a project first went to the people who actually did the research and design, and then it went to a whole different ministry when it came time to fabricate the thing. This was more like an industrial fabrication ministry. So all the way up to the top, the management was somewhat de-coupled in those two processes.

That was kind of the background of the story. So, I asked my friend, "You know, you had those colored paint chips on the Venera landers but I don't understand something because every time I've seen a colored picture of them, I don't really see much variety of colors. There's some orangeish looking things and some whitish looking things. Where are the other colors?" Then he starts laughing and tells me the story.

He said they managed to send Venera there with paint chips that faded in the 900°. And I said, "But surely you knew that. I mean, *you* had discovered that it was 900°." He said, "Oh yes, yes we knew all about that and they developed all these fabulous paints that would withstand the environment of Venus, but that was the design phase." When it got translated over to the other ministry, they were down in some factory manufacturing it and some worker said, "This is ridiculous. We're spending $3,000 on this paint chip here and we can do this with much cheaper paint." And so they did. They went out and they bought cheap paint – after all the development of this high-temperature paint. And he says he thinks the worker got some sort of a Lenin prize or some award for innovation – he was highly rewarded for it."

ABOVE THE RED PLANET

Mars is a difficult place to study. The thin atmosphere makes a Venera-style slow parachute descent impossible. Probes entering from space get little help in slowing down from the frail Martian air, so the window of atmospheric study from the vacuum of space to the ground is narrow. Probes must act quickly and transmit on the go if real-time data is to be culled from the distant world. On March 12, 1974, the Soviet Mars 6 returned the first data from the atmosphere of another world, transmitting data for 224 s before contact was lost at or near the surface.[1] Victor Kerzhanovich was at the Russian Scientific Research Institute at the time. "There was only one Mars probe that actually reached the surface, but no data from the surface was received, or not much. But we got this atmospheric data from the descent.

The Mars 6 lander, carried within its conical aeroshell at the top of the interplanetary cruise bus (© Mark Wade)

It was just pressure and temperature measurements, and again, one-way Doppler. So we combined all of these things to extract my profile of temperature, pressure, and wind on the way down. The parachute did open, but we didn't have a successful landing. Mars has never been a good target for the Russian program. As you know, the first American lander landed on Mars, and I believe it was beyond our capabilities at this time. For Venus, it is much easier because on Venus, the atmosphere is so dense if you didn't miss the atmosphere, you will have some success. On Mars, unfortunately, not so much."

Despite its failure to land, the craft provided insight into the structure of the Martian atmosphere beginning at 82 km and passing through the base of the stratosphere at 25 km above the surface. Near-surface pressure was measured at 6 mbar, or 6/1,000th that of Earth. Temperatures ranged from −189°F in the stratosphere to −45°F near the surface. A mass spectrometer charted gases, but its data was recorded for playback after landing and so was lost. Mars 6 also reported several times more water vapor than was previously deduced from orbiters and flybys.

U.S. plans to land on Mars culminated in the Viking project. For planetary meteorologists, Viking was to be a double shot. The orbiters would study Martian weather from above with instruments more advanced than those of Mariner 9. Even better, Viking landers were to measure the atmosphere in situ during their descent and follow these data with first-hand weather observations from the surface. Viking's aeroshell contained its own mass spectrometer, to detect what kinds of gases were

1. The Soviet Mars 3 made it down to the surface on December 2, 1971, but did not return any useful data.

present in the upper atmosphere, as well as instruments to monitor atmospheric density, pressure, and temperature. A retarding potential analyzer measured ions flowing by the spacecraft. Vikings gave an accurate profile of Mars's atmosphere from about 62 miles in altitude all the way to the surface.

The Viking landers each deployed a meteorology boom. Instruments at the end of the booms sampled the environment from a perch about 4 ft (1.3 m) above the surface. This placed the experiments well within what scientists call the "boundary layer," a region of atmosphere strongly affected by the surface temperature and chemistry. On Mars and other solid-surfaced worlds, it's an important layer for cosmic meteorologists to understand, says MRO's Steve Lee. "Things behave very differently inside that layer than they do even centimeters above the surface. And so that's where a lot of the energy is exchanged between the atmosphere and the surface."

The Viking landers returned beautiful color images such as this one of Utopia Planitia. Note the salmon-colored sky, tinted by the ubiquitous Martian dust (NASA/JPL)

Viking weather stations tracked temperature, pressure, wind direction, and speed. For the first 8 months of surface operations, effects of a subsiding global dust storm dominated meteorology. Late in 1977, things settled down to regular seasonal patterns, giving scientists their first long-term portrait of Martian weather. Temperatures rose and fell as the seasons changed, but so did the planet's pressure. This dramatic relationship between temperature and pressure marks the ebb and flow of the polar caps, which can freeze out and trap substantial amounts of the planet's atmosphere (see Chap. 6).

The Vikings charted other seasonal changes. The landers endured several of the planet's global dust storms. Viking 2, farther north than its sibling, even observed ground frost. The nuclear-powered landers lasted 6 years and 3 months, and 3 years and 7 months, respectively.

Morning on the plains of Utopia brought frost in winter (NASA/JPL)

Pathfinder's rover Sojourner examines the rock "Yogi" at sunset (© Michael Carroll)

Twenty years after Viking, another successful landing finally took place on the rusted plains of Mars. The Pathfinder lander, carrying a small rover, used a series of parachutes, retrorockets, and airbags to land in Ares Valles, a flood plain that appeared to be a catch-basin for rocks from many areas. During descent, the craft measured outside temperatures. Pathfinder carried a small rover the size of a microwave oven. It also had a host of meteorological equipment, including instruments to measure temperature, pressure, and winds. Pathfinder's temperature boom measured temperatures at three levels: 25, 50, and 100 cm above the surface.

Ever since Viking, researchers wanted to know how the temperatures and air flow behaved close to the Martian ground. Peter Smith was on the Pathfinder team. In addition to his work on Phoenix, Smith was Principal Investigator for the IMP imaging system for Pathfinder. "If you were standing on Mars, your head could be 20°C cooler than your feet. Isn't that interesting? I never thought of a place like that. What happens is that the Sun's energy comes straight down to the surface. It doesn't get absorbed much in the atmosphere. The surface gets hot, but there's not much mixing of the air. Convection isn't a very powerful force on Mars, so there tends to be very large temperature gradients. On Earth, you tend to think your head and feet are the same temperature. Not on Mars. You'd need a good hat and a warm jacket but you could walk around barefoot!" That steep temperature change, called a gradient, makes for unstable air near the Martian surface. Pathfinder's wind and pressure sensors confirmed the existence of dust devils firsthand, and even imaged two during the mission.

Dust devils spun across the Pathfinder landing site multiple times (NASA/JPL)

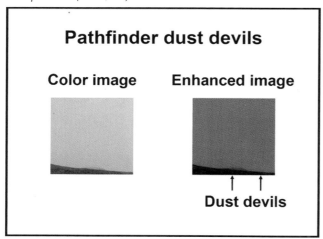

Pathfinder dust devils

Color image   **Enhanced image**

↑   ↑
Dust devils

The Sojourner "microrover" rolled off the lander and wheeled around the immediate vicinity for the entire 90-day mission. The small rover was equipped with a device to test rock types, but the device was often thwarted by the ubiquitous Martian dust. Mars dust is the consistency of talcum powder. It confounded Smith's efforts to see the spectrum of rocks and

Pathfinder views of the Martian sky (NASA/JPL)

soil during the Pathfinder mission. "I was disappointed in some respects. I had put 24 filters on our cameras to try to detect very subtle differences, and dammit if the rocks didn't all look the same. Some were brighter and some were darker, but there really wasn't much in the way of color in the rocks. Maybe if you had a hose up there and you could wash them off, you'd be in business. That dust gets everywhere, and it just sticks all over the rocks. [Our filters] were actually sensitive to chlorophyll and lots of minerals. I was hoping there would be three or four different types of rock. We had landed in a place that we called 'a grab-bag of rocks.' It was at the outflow of Ares Vallis. The idea was that all kinds of different rocks would be carried out of that drainage basin. But all we saw was brighter and darker of the same spectral shape." Because of Sojourner's dusty challenges, later rovers were equipped with steel brushes to clean off target areas on rock surfaces.

Pathfinder lander photo of the Sojourner rover in action (NASA/JPL)

The Pathfinder/Sojourner mission was the product of a new vision and management style for NASA. Not all management experiments are successful. The approach was summed up in the theme slogan "Faster, better, cheaper," remembers Peter Smith. "In the late 1990s, we were under an administration with Dan Goldin that said, 'Let's take great risks.' One way to make sure you are taking risks is to cut your budget, and then, for certain, you'll be taking great risks because you can't afford to do things the proper way. He thought that would engender creativity. He wanted to send lots of inexpensive missions, and some of them were going to fail. The Mars Polar Lander, unfortunately, was a victim of that philosophy."

The Mars Polar Lander (MPL) was due to land on Mars December 3, 1999. It was done on a conservative budget and on a short schedule. The craft was to land in similar fashion to the Vikings, using a parachute and retro rockets. Its mission was to uncover the climate history and volatiles of Mars by literally digging into the south polar region of the Red Planet. The solar-powered lander was equipped with an imaging system, a robotic arm for digging and retrieving soil and ice samples, a gas analyzer, and a meteorology package.

Peter Smith was one of the Principal Investigators. "Our budget had been cut to the point where you just had one engineer working on each subsystem. There was nobody there to check and make sure he or she was doing the right things. It came to the point where you're really risking this one person making a mistake and not catching it; it's very hard to catch your own mistakes. Our navigation team was stretched to the limit, and they were making mistakes. It was really a terrible situation overall." The stretched resources led to failures. First, the Mars Climate Orbiter arrived. A navigational error sent it burning up in the Martian atmosphere. Smith's team felt a sense of panic, but there was little to be done. "You couldn't put it into a parking orbit and figure things out. One way to save money is to remove all the communication during entry, descent, and landing. There was no way to tell what went right and what went wrong. For the next mission, Phoenix, Charles Elachi [Director of JPL] insisted on signals being picked up by two of the Mars orbiters at once, just in case one of the oops happened. He did not want a situation where the lander failed and nobody knew why. That would have been the worst thing in his opinion, and I tend to agree with him. Future missions rely on that information. You want to be able to learn from your failures so you don't have the same one twice."

Analysis would later reveal the likely culprit in MPL's failure – a computer software glitch. When the craft hit the ground, a sensor would shut down the engines. But the sensor may have been activated too early. Rather than sensing the gentle bounce of terra firma, it sensed the bump as the landing gear was deployed at about 10,000 ft. MPL may have been in freefall for the rest of its descent. In short, the Mars Polar Lander was done on the cheap by capable people lacking necessary resources. JPL management wanted to prevent a rerun. Its follow-up mission to the Mars Polar Lander, Phoenix, would not be ready for nearly a decade. In the interim, a trusted, simple landing system would be tried again.

The Mars Exploration Rovers (MER) used similar technology to Pathfinder's for their landings on Mars in January of 2004, but the Pathfinder parachute and airbags had to be beefed up considerably for the larger craft. Each MER weighed in at 1,175 lb, a significant increase over Pathfinder/Sojourner's 604 lb. NASA pundits call the rovers "robot geologists,"

Artist's concept of the Mars Polar Lander on the south pole of Mars. In reality, the scene probably more resembles a debris field (© Michael Carroll)

Panoramas from Spirit in the Columbia Hills (top) and Opportunity at Victoria Crater (NASA/JPL)

as their primary mission was geology-centered. However, the rovers were also set up to monitor weather visually, and their Miniature Thermal Emission Spectrometers (Mini-TES) could provide meteorologists with temperature profiles of the atmosphere above.

The rovers found that surface heating began to disturb the air at about 10 a.m. local Mars time. Ground temperatures are 20° warmer than the air above. Spirit and Opportunity carried out "long ground stares," looking at one section of surface for 8½ min sessions. These sessions showed about 10° temperature changes as packets of warm and cool air moved across the Mini-TES field of view. The temperature data looking up enabled scientists to discern water vapor levels and temperatures at different altitudes. The rovers have observed dust storms and dust devils, clouds, and light scattering at sunset.

At the same Mars launch window, the European Space Agency sent its Mars Express. The orbiter carried a British-built probe called Beagle 2, ESA's first Mars lander. Beagle 2 was the brainchild of Professor Collin Pillinger of the Open University. At 25 in. diameter, it was the smallest, most heavily instrumented lander to date. The craft was to land on dual airbags similar to those used on Pathfinder and Soviet Luna moon landing probes. Beagle 2 carried multiple instruments to search for organic material, including a self-propelled "mole" to dig samples and a sampling arm to take soil and rock to a mass spectrometer. A gas analysis package would have checked atmospheric and soil results against those from Viking.

ESA's Mars Express orbiter televised the departure of Beagle 2 for a landing on December 3, 2003. The lander disappeared somewhere in the Isidis Planitia region. Sadly, contact with the ingenious probe was never established, despite international efforts using several ground stations and Mars orbiters. As with so many Mars probes, the cause of the mission failure remains a mystery.

Beagle 2 would have deployed four solar panels and a "PAW" to investigate the soil in search of biologically oriented processes (All rights reserved Beagle 2, http://www.beagle2.com/)

For the United States, the Pathfinder and Mars Exploration Rovers were a stunning success, but they reached the Martian surface using a landing system that could not be used for larger spacecraft. In fact, engineers felt that the MERs were at the very upper limit of airbag technology. Additionally, systems were on the drawing board for far more delicate instrumentation that would require a gentler landing. The next Mars lander would need to utilize the same kind of landing technology that had not been used since Viking, landing with parachutes and steerable rocket engines.

It was up to the Phoenix lander to prove a soft landing could be done, and to explore polar terrain that the doomed Mars Polar Lander tried to explore nearly a decade before. Phoenix, named after the mythical bird resurrected from the ashes, was aptly named, as the craft flew many experiments originally on the lost MPL.

JPL's Charles Elachi got his wish. The landing of Phoenix was monitored by multiple ground stations and by three Mars orbiters, NASA's Mars Odyssey, the Mars Reconnaissance Orbiter, and ESA's Mars Express. The Mars Reconnaissance Orbiter was able to image Phoenix during its parachute descent.

Phoenix imaged water ice under its main bus, exposed by the exhaust of its engines upon landing. It also found ice near the surface in trenches dug by its robotic arm. Polygonal ground at the landing site shows the same characteristics that tundra areas do on Earth. Phoenix's Canadian meteorology package provided detailed arctic weather reports throughout the lander's 5-month mission. Its successful landing paves the way for larger, more sophisticated landers in the future. (For more on Phoenix's results, see Chap. 6.)

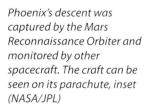

Phoenix's descent was captured by the Mars Reconnaissance Orbiter and monitored by other spacecraft. The craft can be seen on its parachute, inset (NASA/JPL)

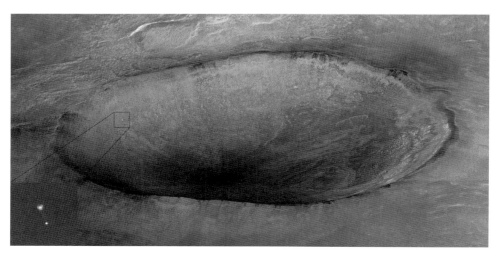

The surface can inform meteorologists of atmospheric conditions. Compare the patterned ground at the Phoenix landing site in the Martian arctic (bottom) to (top to bottom center) Viking 1, Viking 2, and Pathfinder/Sojourner sites. Because of a combination of temperature, pressure, and condensation of carbon dioxide and water, the northern arctic soil at the Phoenix site is saturated by water ice, forming domed polygons across the tundra-like landscape (Top to bottom: NASA/JPL, NASA/JPL, NASA/JPL, NASA/JPL-Caltech/University of Arizona/Texas A&M/James Canvin-www.nivnac.co.uk/mer)

These images may show briny water droplets forming on the legs of the Phoenix lander. Note the two drops that seem to coalesce (NASA/JPL/UA)

Martian Argon: Clue to Past Atmosphere?

Noble gases (gases that do not combine with others, including xenon, argon, and neon) provide a good yardstick to estimate how much atmosphere a planet had in its beginnings. As other gases such as helium and hydrogen are lost to space over eons, noble gas levels remain the same. The more atmosphere a planet has to begin with, the more noble gas is left behind. This makes the measurement of argon levels in an atmosphere very important to our understanding of a planet's early atmosphere and weather. Argon levels on Mars were a holy grail that atmosphericists were seeking when probes first began sampling Mars atmosphere in situ. The earliest attempts came from Soviet landers. The first to return data about argon levels was the Soviet Mars 6 in March of 1974, says veteran Mars geochemist Ben Clark. "There was an interesting controversy. The Russians had claimed a landing, but that landing was questionable. They got 30 s of white data. But on the way down they had something like a mass spectrometer with a pump, and from the data they got, they concluded that the atmosphere must have a lot of argon in it. They thought the argon was up around 20 or 30%. That would be really important because that would be an indication of the possibility that there was indeed a much thicker atmosphere of the other gases at one time, and then argon had remained when the other gases had been stripped away." Many in the scientific community were skeptical about the readings and anxiously awaiting a second opinion, which would come from the U.S. Viking landers in 1976. "When we landed, we had a contest on the amount of argon we would find, and the person who won the contest was an engineer, because he didn't understand the scientific literature that said everybody knew it had to be at least 20%. The Russians were wrong. It turned out that my instrument [a laser-based soil analyzer] could measure atmospheric argon. We'd shoot the X-rays into the soil, but it went across a distance of about an inch. Then, of course, after it hit the soil it came back again, and as it crossed that distance, the argon got excited and sent its signal back as well. So we did a measurement with our system and we were able to put a much lower value on the argon. There was also a mass spectrometer [aboard the Vikings] that measured it during the entry."

Thanks to the many landers and atmospheric probes that have studied Mars, its level of argon is now known to be 1.6%. Clark is able to use equipment aboard the Mars Exploration rovers to measure argon in similar ways. Scientists hope to further refine argon levels to better understand conditions on ancient Mars. "Today on Mars we're taking argon measurements every few weeks with a similar instrument that measures the soil. We're measuring that as sort of a poor man's pressure gauge."

THE OUTER PLANETS: GALILEO-JUPITER, HUYGENS-TITAN

To date, only two atmospheric probes have visited the outer Solar System. The Galileo probe parachuted into the atmosphere of Jupiter on December 7, 1995. ESA's Huygens probe made landfall on Saturn's moon Titan nearly a decade later, on January 14, 2005.

The Jovian environment offered scientists a time machine of sorts, a way of looking at the building blocks of the early Solar System. Jupiter is the king of the worlds, largest and heaviest. If all the planets and moons in our Solar System were combined, they would still fall short of Jupiter's mass. Jupiter seems to be the poster child of the outer planets. It is the most accessible, so scientists wanted to look at it as the prototypical gas giant, learning about Jupiter's distant siblings (Saturn, Uranus, and Neptune) by association. Over the course of 4.6 billion years, the inner Solar System has evolved and changed into a set of planets quite different from their primordial natures. Researchers reasoned that Jupiter formed far enough away from the Sun's heat and solar wind to retain much of its original makeup. Galileo was to discover what constituents made up Jupiter's atmosphere, how abundant each was, and how that atmosphere was structured. Relative abundances tell an important historic story. For example, the amount of water within Jupiter's clouds could give scientists insights into how much oxygen was involved in the planet's formation, and how much oxygen was present in the early Solar System. As we saw at Mars, noble gases such as argon and xenon afford scientists valuable information about the evolution of atmospheres. These cannot be seen from orbiters, but must be sampled in situ by a probe such as Galileo.

More generally, how does a planet behave when it has no solid surface? Jupiter's countenance stands in stark contrast to Earth's. It is ten times as far across, and it spins at breakneck speed, taking only hours for a complete day. Earth's mid-latitude clouds tend toward cyclonic storms or calmer spirals, but Jupiter's clouds are stretched into belts (fairly clear, dark stripes) and zones (bright bands of cloud). What could we learn about Earth's complex storms, airflow, atmospheric chemistry, and climate by looking at a weather world? To terrestrial and space meteorologists, as well as other planetary scientists, it was clear that Jupiter was an important place to go.

Galileo's orbiter, with its crippled main antenna (see Chap. 3), had to transmit data to Earth at severely constrained rates. Fortunately, the probe had a fairly low data rate that would be unaffected by the failure. But due to the Challenger disaster, Galileo's engineers had to design a new, circuitous path through the inner Solar System, extending flight time to Jupiter by years. The probe had been designed for a 3-year flight. It now had to endure a cruise twice that long. And after being chilled, toasted, juggled, and exposed to vacuum of space, Galileo would endure the harshest entry conditions in the entire Solar System. Would it survive to tell tales of distant lands?

The Galileo probe bristled with a host of atmospheric instruments (NASA/JPL)

The orbiter released the probe 148 days out from Jupiter (close to the probe's battery lifetime of 150 days). The early release gave the orbiter enough separation from the probe to safely miss Jupiter and go into orbit. The probe systems were controlled by a simple timer, and powered up some 6 h before entry. Entry began at 280 miles above the atmospheric level at which the pressure was 1 bar (equivalent to Earth's surface pressure). Entering Jupiter's atmosphere on December 7, 1995, the Galileo probe survived entry speeds of over 106,000 mph, temperatures twice those on the surface of the Sun, and deceleration forces up to 230 times the strength of gravity on Earth. The probe deployed a small drogue parachute, which pulled the back shell from the probe and dragged out the main parachute. Ten seconds later, Galileo dropped its protective heat shield, called

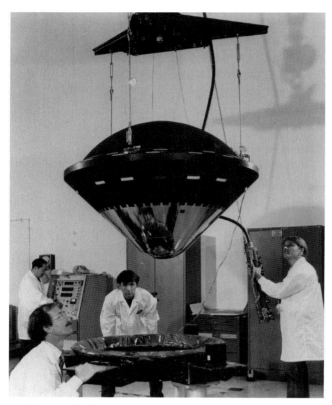

Galileo rested within a carbon phenolic heat shield that could survive temperatures hotter than the surface of the Sun (NASA/JPL)

an aeroshell. About 1½ s later, with the aeroshell 30 yards away, the official mission commenced, and Galileo began to investigate the dramatic skies of Jupiter. It relayed data obtained during its 57-min descent mission back to the Galileo orbiter more than 130,000 miles overhead for storage and transmission to Earth (real-time data had to be stored because of the orbiter's failed main antenna).

Engineering data from the probe indicates that it had a rough ride. Galileo began transmitting 53 s late due to a wiring problem. The probe swung back and forth beneath its parachute every 5 s and was apparently spinning once each 20–25 s. Winds at 400 miles per hour buffeted the lonely emissary as it fell through canyons of clouds and smoggy hazes.

In addition to water and other constituents, Galileo was charged with monitoring ammonia, according to planetary researcher Kevin Baines of NASA's Jet Propulsion Laboratory. "Nobody can spectroscopically identify ammonia on either Jupiter or Saturn, but we did it with Galileo. Ammonia clouds are associated with dramatic updrafts in active areas, and they are freshly formed. That's where we saw the most predominant ammonia clouds. We also saw them associated with the hot spots." But Jupiter threw another curve at Galileo, Baines says. "You send one probe to Jupiter and you want to get an average place, but it fell into the most un-average place on the whole planet and the driest place, a hot spot where there are no clouds. It was all dictated by celestial mechanics. The probe had to go in at a certain latitude, and whatever was there longitudinally was what you got. It just happened to fall into the brightest hot spot on the planet at the time."

For a few days, scientists struggled to understand what had happened. The probe had passed through very few high-altitude ammonia clouds, and even fewer water clouds. It was beginning to look as if all the models of Solar System evolution were wrong. Baines and his team knew that the amount of water in the air provides insight into how much oxygen there is in the Jovian system, and oxygen is a gas that helps scientists reconstruct the history of the atmosphere. "If there's no water, that means there's no oxygen. Then we realized it fell into a hot spot. You cannot say there's no oxygen; it just went into a dry place. We didn't get the information on the water content or, more specifically, the oxygen content, of the planet. That was one of the big things we were going to do, a primary goal. But from orbit we were able to prove that there was more water elsewhere. If that probe had just fallen 3,000 km

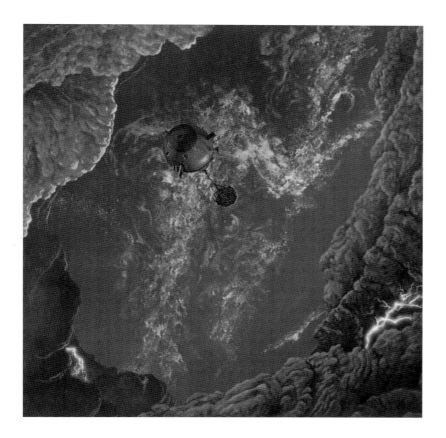

This artist's concept reflects what scientists hoped Galileo would experience. But the probe sailed through uncharacteristically clear skies (© Michael Carroll)

away (and 3,000 km is a very small distance on Jupiter) it would have found lots of water."

One thing the probe did find was heat. Jupiter gives off more heat than it receives from the Sun, and Galileo's net flux radiometer experiment helped fill in the details. Jovian heat is involved with many mysteries, including the relative abundances of gases and processes taking place deep within the planet. Goddard Space Flight Center's Anthony Delgenio outlines those deep processes. "You have the heat left over from the formation of the Solar System, but then you also have the additional heat that's been produced by helium precipitation. As the planet has aged and cooled, helium has gradually separated out from the hydrogen and begun to precipitate out toward the center. The kinetic energy associated with the falling helium droplets then gets converted into thermal energy. That helium precipitation is then another source of internal heating."

As the probe continued to gather its precious data, internal temperatures increased to dangerous levels. During its fiery entry, technicians believe hot gas entered into the shell around the probe, perhaps through a breached seal. Experiments were calibrated for certain temperature ranges, and the probe's temperature quickly rose above those levels. This created challenges for scientists who tried to interpret the numbers coming back from Jupiter. Nevertheless, Galileo's legacy remains a turning point in our understanding of planetary atmospheres and weather (see Chap. 7).

Jupiter is 5 times as far from the Sun as Earth is, but Galileo was not the most distant probe to sample alien weather. A plucky European probe called Huygens holds that distinction. Carried into the wintry outer Solar System by NASA's Cassini Saturn orbiter, Huygens arrived at the upper fringes of Titan's air on January 14, 2005. The probe nestled within a 9-foot diameter heat shield. Because of Titan's low gravity (a tenth that of Earth's) and dense atmosphere, the probe required much less protection than that of Galileo, says biophysicist Ben Clark. "The atmosphere is extended very high because of Titan's low gravity, and that makes Titan unusually easy to land on. Parachutes work great. Entry velocities are low." In fact, engineers estimated that Huygens would enter the atmosphere at a speed of 13,000 miles per hour, experiencing 12 G's of deceleration (compared to Galileo's 350 G's).

As moons go, Titan is a behemoth, spanning a diameter larger than the planet Mercury. The first flybys by the Pioneer and Voyager spacecraft revealed an opaque orange atmosphere and nothing of the surface. But scientists gathered enough information to realize that complex chemistry was taking place on the mini-planet, chemistry that might lead to lakes of methane and prebiotic conditions similar to the primordial Earth. University of Colorado's Nick Schneider remembers the excitement of discovery at Saturn's moon. "Titan was pretty much a blank slate until Voyager. Everybody knew that it had nitrogen in its atmosphere and that it was cloudy. But beyond that? You should see the space art for Titan. It got even wilder after Voyager. The reason is that Voyager discovered the chemical composition of the atmosphere in greater detail and determined that hydrocarbons should be accumulating. There was a pivotal paper by Jonathan Lunine that postulated oceans of ethane; perhaps the whole surface was liquid. Even nowadays it's pretty wild, and justifiably so. The current story is methane lakes with a little ethane mixed in, so everything we've got here for a hydrological cycle on Earth has its equivalent for methane on Titan."

And with the Hubble Space Telescope, scientists were finally able to see faint features on the surface, says Peter Smith of the Lunar and Planetary Laboratory. "With Titan I managed to get some Hubble Space Telescope time. I used a special filter that I could see to the surface with. There's a little window out at about 950 nm (light wavelength) that enables you to see down to the surface. We didn't know that ahead of time, but I wrote a proposal and said, 'I think we can see to the surface at this wavelength. I want to use this particular filter to map the surface, and at the same time we want to map clouds, too.' So the reviewers said, 'No way on Earth you're going to see to the surface, but we'll let you map the clouds.' They cut half my time, and of course we saw no clouds but made a beautiful map of the surface. The thinking at that time was that you couldn't see to the surface, because of the design of Voyager. But Voyager's longest wavelength was something like 650 nm, and they just didn't have a long enough filter."

With Hubble's new data, flight planners had something to shoot at, says aerospace engineer Pat Carroll. "Every new bit of data changes the approach the scientists can take. And eventually you get enough data to decide where you'd like to go specifically."

Huygens was a spectacular success. Peter Smith helped to develop the Descent Imager/Spectral Radiometer. "I became the project manager, which meant flying up to Lockheed/Martin all the time and managing their progress in designing the imager. While we were flying up there we'd look out the window and see all these wonderful features. So we'd think, 'Okay, imagine we're hovering over Titan.

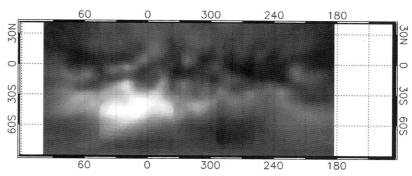

At some wavelengths, the Hubble Space Telescope was able to discern subtle features on the surface of Titan (ESO)

Our camera hasn't got this kind of resolution. It's more like if you take your glasses off.' So we'd take our glasses off and look out the window. We'd know the reality because we'd just seen it, and we would say, 'Oh, this isn't good. Can't see a damn thing!' So we'd put our glasses back on and sorta squint so we couldn't see as much, but it wasn't as bad as seeing without glasses, and that's what we went for in our camera. We were thrilled with the results."

Smith's DISR imager scanned through a thin slot. Huygens was designed to spin, so images from the DISR could be assembled into 360° panoramas. Huygens returned spectacular images from high altitude as well as on the surface. "For those first images of the surface, I was back in Germany with the inner circle of people who first saw these images, and they were just astounding. It looked like the coast of Italy: streams coming down a hill into what looked like a dry lake bed and things that looked like a fault line that had shifted. You could see a straight line and a sort of island that had been pushed away. It just looked awfully familiar. Of course we know that the rain is methane and the rocks are water ice, and there are hydrocarbons all over the place, but still, the shapes of the surface looked awfully familiar. That was a big surprise to me."

Smith notes that the mission had another, unwanted surprise: the probe did not perform the way it was designed to in flight. "It spun, all right, but it spun in the opposite direction to the one they told us. That would have been fine, except we had built the instrument to work spinning counter clockwise, and we were timing off signals from the Sun sensor. Every time the Sun went by we got a pulse that triggered our timing. We were trying to get spectra of the sky near the Sun, polarization spectrum. We were really studying the atmosphere as much as the surface, because for most of the descent you can't see the

Huygens full-scale test article. Note the thin wind vanes covering the probe's sides (Wikimedia photo by David Monniaux)

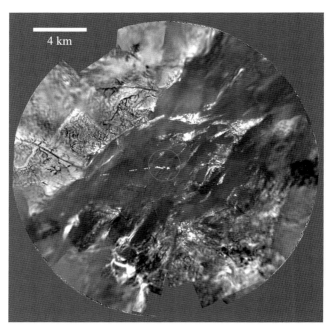

Huygens mosaic taken from 11 to 5 mile altitudes was at once alien and familiar (JPL/University of Arizona)

surface. Our spectral radiometers were tuned for understanding all the optical properties of the haze. Since we spun the wrong way, it was quite a difficult challenge to decode the data. [Principal Investigator Marty Tomasco] spent an amazing effort, and I think he got about as much as he hoped he could get."

Pioneer engineer Pat Carroll, who came up with some of the first designs for Titan probes even before the Pioneer 11 encounter, was pleasantly surprised, too. "There was so much we didn't have details on. One of the surprises was how well the cameras worked. There was plenty of light to see detail. We didn't know much about the light levels. We were excited about it, about the details and the ability to collect this kind of data. It's pretty amazing." (For Huygens' results, see Chap. 8).

A Whiff of Titan: How Huygens Did It

ESA's Huygens Titan probe carried a sophisticated battery of instruments to study the structure and composition of the atmosphere, to chart the winds, and to visually survey the cloud layers and surface. The science package consisted of these experiments:

The *Aerosol Collector and Pyrolyser* (ACP) sampled fine mists as Huygens descended through various cloud layers. The device deployed a snorkel at two separate times during the probe's journey through the atmosphere. The first samples were taken from the top of the atmosphere down to about 40 km altitude. The second collection came from between 23 and 17 km. Each time, a pump pulled air through a filter to capture any droplets. At the end of each sampling period, the filter was pulled into the probe. There, the GCMS (see below) analyzed the sample at ambient temperature. Then, a furnace heated samples at 250°C, then at 600°C. The different temperatures revealed different constituents in the atmosphere.

The *Descent Imager/Spectral Radiometer* (DISR) was a complex imaging system that included photometers,

visible and infrared spectrometers, a side-looking imager, two downward-looking imagers (medium and high resolution), and a solar aureole sensor. As the probe spun, DISR scanned up and down, assembling panoramas of the landscape below.

The *Doppler Wind Experiment* (DWE) used the carrier signal on Huygens to chart changes in the probe's acceleration and drift in the atmosphere. DWE also afforded researchers a picture of the probe's motions under its parachute, along with some data on atmospheric layers above the descending craft.

The *Gas Chromatograph and Mass Spectrometer* (GCMS) analyzed various elements in the atmosphere, along with complex organic molecules.

The *Huygens Atmosphere Structure Instrument* (HASI) sensed the physical properties of the air, electrical charges, and sounds.

The *Surface Science Package* (SSP) investigated the consistency and temperature of the soil, and the dielectric constant of moisture in the surface material.

Part II

The Forecast: Clearing with Scattered Ammonia Showers by Morning

The ruins of Uxmal serve as a reminder of the important role that Venus played in the pantheon of ancient Mesoamerica. Its preeminence in the night sky and its 225-day orbit – similar to the gestation period of humans – contributed to Venus's status as a link between the affairs of humans and the heavens (© Michael Carroll)

Chapter 5
Venus

Across eastern Mesoamerica, Mayan priests scaled precipitous stairways to temple summits. As clerics had done for half a millennium, they reached a platform to begin ceremonies dedicated to the god of the evening star, Hunahpu.[1] Draped in jaguar skins, their painted faces looking skyward beneath a headdress of quetzal feathers, the priests made their way to sacrificial altars. New sacrifices, both animal and human, brought a fresh blaze of crimson to ancient bloodstains. It was a crucial time, a moment in the cosmic travels when Hunahpu disappeared to visit the underworld, Xibalba. There, the god of the morning star battled the forces of darkness on behalf of humanity. Later, it was hoped, he would rise as his divine brother, Xbalanque. Would he be successful? Would he return to make all things new again on Earth? Only the prayers of the people and the sacrifices of the high priests could assure Hunahpu's resurrection.

The veneration given Venus by the ancients only makes sense. Brightest of the "wanderers" in the sky, Venus never strays far from the life-giving Sun. It moves in predictable ways that cycle through a period of time similar to the gestation period of humans (its path takes it around the Sun once each 224.7 days). The planet's looping motion across the sky follows five specific movements every 8 years. Its pentagonal celestial voyage was understood and well documented by the Mayas. Mayan priest-kings assigned symbols to each cycle.

The Mayans were not the first to revere the evening star. To the Babylonians, the bright planet conjured up associations with their god Ishtar. The Sumerians worshipped Innana, and the Etruscans worshipped the planet in the form of their bejeweled goddess Turan. Later, in the northern Americas, the Skidi Pawnee sacrificed humans to Cu-piritta-ka.[2]

For the western world, the most familiar deity associated with Venus is Aphrodite, goddess of beauty, love, grace, and sexual exhilaration. Would the Greeks have named the planet after this graceful goddess had they known of its true environment? The Venusian landscape simmers beneath oppressive air ninety times the density of Earth's, equivalent to the pressure a kilometer deep in the ocean. Average surface temperatures reach a searing 864°F. At night, the rocks are so hot that the Venusian landscape glows a dull red. Sulfuric acid rains from an eternally overcast and gloomy sky. But perhaps the Greeks were on the right track. Aphrodite (Venus is the Roman version) was seen as tempestuous, sultry, hot-tempered, and alluringly beautiful. Venus is the brightest planet in the twilight sky. Through the telescope, her face glows as a perfect white orb. But her skin-deep beauty belies her Dantean environment.

In physical character, Venus is Earth's twin sister. She is essentially the same size, only slightly less dense (95% that of Earth), and probably developed in the same solar neighborhood, being the closest planet to us. Like Earth, Venus has volcanoes that must have contributed water vapor and gases to its primordial atmosphere. In fact, fully 90% of Venus's surface has been affected

1. The plumed serpent, also known by the Nahuatl name Quetzalcoatl.

2. Literally the "female white star."

by volcanic processes. As on Earth, these volcanoes are powered by the planet's internal heat. Some of the heat is left over from the initial creation of the planet, while other heat is radiogenic, coming from the breakdown of radioactive materials in the core. Venus and Earth and – to a lesser extent, Mars – are large enough to have retained a great deal of their initial heat energy, and their cores create prodigious amounts from radioactivity. The heat must escape, and it does so most dramatically through volcanoes.

Volcanoes play an important role in the weather of a planet, because they contribute to the atmosphere. Carbon dioxide, carbon monoxide, and water vapor pour into the skies of Earth each day. Over 600 volcanoes are currently active or dormant worldwide. Their eruptions spew acidic vapors into the upper atmosphere, causing acid rain. Volcanic ash, carried aloft and suspended in the upper troposphere and lower stratosphere, can lower global temperatures to the point where the climate actually changes.

On our early Earth, the initial atmosphere consisted of matter pulled in from the clouds of the Solar System's creation. This atmosphere was lost to space by simply leaking away or by the massive impacts from asteroids and comets. Volcanic activity and other processes expelled a second atmosphere, primarily of water vapor. The water condensed and the first rains fell, triggering the water cycle. Oceans grew in the lowlands, chemically locking some carbon dioxide into the rocks as carbonates. Sunlight split some of it into hydrogen and oxygen. These gases combined with others to form the air we have today, transformed by living systems (see Chap. 1).

Venus suffered a different fate. Instead of a thin nitrogen/oxygen atmosphere, a dense canopy of almost pure carbon dioxide blankets the planet next door. The different paths taken by the two cosmic siblings are due, in

Topography of Venus and Earth compared. Where terrestrial lowlands filled with oceans (purple and blue regions), Venusian plains remained parched, its rain never making it to the surface (Left: NASA/JPL/NOAA. Right: NASA/JPL)

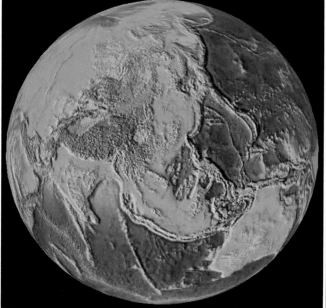

part, to Venus's proximity to the Sun. Venus is 7/10 as far from the Sun as Earth is. Its atmosphere captured and held the heat from the Sun. Just as car windows let in sunlight, convert it to heat energy, and trap it inside long enough to melt your sunglasses to the dashboard, Venus' dense atmosphere blankets it in a heat-trapping cocoon, ensnaring the heat in a self-reinforcing greenhouse cycle. As volcanoes built its water vapor atmosphere, the air was too hot for the water to condense. No rain fell. No

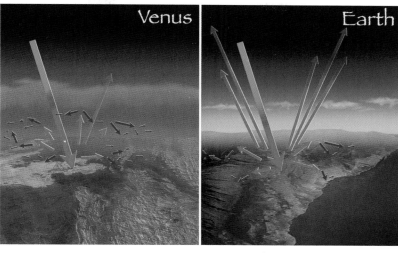

The greenhouse effect compared: Venus's atmosphere holds most of the Sun's energy as heat, whereas Earth allows more heat to escape back into space

oceans settled in the basins. Instead, carbon dioxide levels steadily rose. Water vapor molecules leaked into space or chemically combined with other gases to make a nasty brew of sulfuric acid in the clouds.

Lockheed/Martin's Ben Clark contrasts the CO_2 on the two cosmic neighbors. "Earth has a small amount of carbon dioxide in the atmosphere, but if you could take all the CO_2 back out of the limestone on Earth, you would have an atmosphere that's about as thick as Venus's atmosphere. One thought about why Venus has this greenhouse problem is that its carbon dioxide never got taken up and sequestered." According to some estimates, the carbon dioxide in the Venusian atmosphere is roughly equal to the amount of carbon in Earth's air and in its carbonate rocks. So, Venus has the same kind of atmosphere that our world would if our carbon had not been shunted away into the rocks by their interactions with our oceans and air. For this reason, environmentalists are very interested in reducing humankind's contributions of carbon dioxide into the air.

With all of its water vapor, cloud cover, free-floating carbon dioxide, and sloshing oceans, our planet would seem to be prone to a runaway greenhouse effect, the same that befell Venus. But Earth's atmosphere has a self-correcting system for balancing out temperatures above and below its heat-trapping clouds. As temperatures rise, water evaporates. This water vapor condenses as clouds. Our clouds reflect more of the Sun's energy back into space, cooling the surface. The clouds act like a thermostat in a home, kicking in their cooling effects just as things start to get hot.

APHRODITE'S AMBIENCE

The weather on Venus has bewildered observers since the first telescopes focused in on the planet. It has long been known that Venus is covered in a thick layer of clouds. But far from the carboniferous swamps envisioned by early academics (see Chap. 3), Venusian clouds are not watery mists.

Venusian meteorology is ruled not by water but by battery acid. High above the visible cloud deck, ultraviolet light from the Sun tears molecules apart in a process called photodissociation. The leftover bits of incomplete molecules frantically try to combine with other things such as sulfur and water to make themselves chemically stable. Venus seems endowed with more than its share of sulfur, and photodissociation forces that sulfur to combine with other elements to form complex effects such as sulfuric acid hazes. It's an exaggerated caricature of sunlight turning Earth's pollution into that famous brown cloud of smog, which ultimately leads to forest- and ocean-damaging acid rain. But Earth's rain is a mere shadow of the strong stuff that comes out of the Venusian clouds; acid rain is a real issue on Venus.

The acidic hazes drift above the uppermost cloud deck, which tops out at about 43 miles above the ground. These visible clouds define the "surface" of Venus that we can see from outside and are also composed of sulfuric acid. The high-altitude haze above condenses into tiny droplets, which fall and congregate into clouds in the warmer air below. The corrosive droplets fall for about 12 miles, growing and becoming heavier. But these clouds are just the beginning: Venus actually has three layers of clouds, each separated by clear air. As the droplets descend, temperatures rise to about 200°F. The sulfuric acid begins to boil, and the cloud deck vanishes into vapor. The sulfuric acid breaks apart in the heat, turning back into sulfur and water. Currents in the air carry some of these chemicals back aloft, where they recombine into sulfuric acid and fall back down again. But a little farther down lies a zone of relatively clear air, separating the upper cloud deck from the middle one.

The second deck stretches from about 32 to 35 miles above the ground. Both the upper and middle cloud decks are made up of particles that range from the size of smoke (called "Mode 1" particles) to the size of water droplets in fine fog on Earth ("Mode 2"). The lowest, third cloud deck is dominated by larger particles, the same size as water drops found in clouds on Earth. These are known as "Mode 3" particles.

The fact that the cloud decks vary in particle size implies that they are made of differing substances. And although sulfur dominates the environment at Venus, the Soviet Venera 12, which actually sampled the clouds on its way to a landing, detected mass quantities of chlorine, twenty times more than sulfur. It is possible that the Mode 3 droplets are chlorine coated with sulfur. We see this "coating" effect at the gas giants. Other theories include some kind of unknown crystalline material floating in the air.

Whatever it is, something is serving as an ultraviolet darkening agent in the clouds. One researcher who is intrigued by the phenomenon is atmospheric scientist Sanjay Limaye, director for the Office of Space Science Education at the University of Wisconsin. Limaye is a member of the Venus Express team's Venus Monitoring Camera. "It's possible that the Mode 3 particles are crystals, like a polymer chain or something. It's unclear. We haven't flown any instrument to sample the particles and actually to look at them. And so the lead tool we have is to look at the scattering properties. And unfortunately this is not a unique process to infer particle properties from phase-angled data. Especially when you have to wait a long time to obtain the phase angled coverage because it's almost impossible for us to observe a given location at all different phase angles." In other words, to study the nature of the mystery darkening agent by observing the way it scatters light is made difficult because of the movement of the spacecraft. That kind of experiment is best done at consistent angles rather than at constantly changing ones. All the previous attempts to decipher the properties of the Mode 3 particles have been only moderately successful, Limaye says. "For example, there is still not a single model that tells you how the properties of a dark area are different vs. a UV bright area. We have no idea."

Larry Esposito is part of Venus Express's Venus Monitoring Camera team. His group is also baffled. "In the data that we have analyzed we're not finding anything yet about the sizes of the particles. But that could happen. If so, we'll write a paper. But right now we're trying to understand more just temporal variations in the brightness, comparing Venus now to Venus in Pioneer Venus days and comparing our results with the camera to infrared results from Venus Express as well."

In situ analysis of the darkening agent would clear up the issue, but sampling this region of the Venusian atmosphere is difficult for an incoming probe, Limaye says. "It certainly would be easier to just go there and sample it. The trouble is when you send a probe we can't sample the top of the cloud layer because most of the instruments don't work until about 52 km above the surface. And the cloud top goes all the way up to 73 km. So that's a very difficult altitude range to sample. Venus is an extremely challenging environment to get data from." For now, the nature of the Mode 3 particles remains a mystery.

At the base of the lowest cloud deck, 30 miles above the surface, the air clears. Into this clear air pours sulfuric acid rain, but it doesn't make it to the ground. High temperatures and updrafts assure that the rainfall on Venus becomes virga, rain that evaporates before it makes landfall. Any inhabitants on the surface would be grateful that the Venusian super-acid rain doesn't make it all the way down.

In the deep, clear air, crushing pressures team with searing temperatures to form a region of thermochemistry, chemistry powered by heat. Here, the ubiquitous sulfur atoms are passed quickly from one nasty compound to another. Chemical processes happen quickly because the high temperatures shove the molecules into a frenzy of movement. High pressures force them together, so colliding molecules constantly break apart and recombine into other compounds. It's a chemistry professor's dream, and there's more chemical fun to come as air meets ground.

At the boundary of the atmosphere and surface is a buffer zone. In this super-heated region, atoms move easily from rock to air and back again. The substances in the rocks actually act as a buffer, calming and controlling the composition of the atmosphere.

While complex feedback loops are circling between the ground and the buffer zone, other balancing interactions are going on further up.

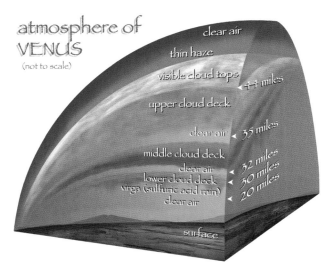

atmosphere of
VENUS
(not to scale)

clear air
thin haze
visible cloud tops
44 miles
upper cloud deck
clear air
35 miles
middle cloud deck
clear air
32 miles
lower cloud deck
30 miles
virga (sulfuric acid rain)
20 miles
clear air
surface

The complex layers of Venus's clouds create truly alien meteorology (© Michael Carroll)

The ultraviolet sunlight at high altitudes is busy creating sulfuric acid, and acid hazes are modulating how much of that same light gets through to lower regions. Clouds and haze layers dictate how much sunlight gets through – and where. Because ultraviolet light and the Sun's heat get through unevenly, being blocked here and seeping deep there, air currents become turbulent. Winds push clouds around, and the atmosphere's superrotation pulls them into patterns and powerful jetstreams. These clouds, in turn, determine where sunlight gets absorbed to make more hazes or clear away clouds. This atmospheric carousel turns at the hands of many complex feedback loops and cycles. The Venusian atmosphere, once thought to be a sluggish ocean of air, turns out to be much more dynamic, says NASA's Victor Kerzhanovich. Kerzhanovich, who was a Soviet scientist with both the Russian Scientific Research Institute and the Institute of Space Research (IKI), says, "This shows how the atmosphere of

Nephelometers aboard Venera and Pioneer Venus probes have returned varying profiles of Venusian cloud decks, providing a glimpse into just how active Venusian weather is (© Michael Carroll, with data from Institute for Space Research (Venera) and NASA/Ames (Pioneer))

Sulphur Cycle?

The Earth has its water, oxygen, and carbon (rock) cycles, constantly reprocessing and renewing materials and gases. Hellish Venus appears to have a sulfur cycle. Although details are yet to be confirmed, circumstantial evidence points to a cycling of sulfur between the air, clouds, and surface, perhaps including outgassing from deep within the planet's subsurface. It all begins in those high-altitude hazes, with sunlight tearing apart molecules in photodissociation. Here, sulfur combines with water and other chemicals, giving birth to massive quantities of sulfuric acid. This corrosive brew forms the basis for all the haze, clouds, and rain on our sister world. At ground level, sulfur is probably leaking out of the rocks, combining with gases such as CO_2 to form even more exotic mixes. Sulfur may even be expelled in more dramatic style by active volcanoes. In fact, the clouds of Venus may have their genesis not in the high hazes but in the rocky soil of Venus. Venusian sulfur, it seems, is on a journey of millions of years, migrating from magmas deep underground to surface rocks, up into the gaseous atmosphere, and back again as vapor and virga (precipitation that never reaches the ground).

Venus is variable. The common idea was that Venus was stagnated, and nothing happens there. But when you go inside, it turned out a lot of things were happening."

INTO THE DEEP

Somewhere beneath the bottom of the lowest cloud deck, within the acidic virga, something strange is happening. The phenomenon – called the Pioneer Venus 12.5-km anomaly – visited all four probes in the Pioneer Venus multi-probe mission, and it baffles analysts to this day. At an altitude of about 12 km (7.5 miles), a power spike surged through all four probes, despite the fact that they were thousands of miles from each other, some in daylight and one in the darkness of night. The surge was accompanied by bizarre readings of temperature and pressure. Many of the instruments failed completely.

In 1995, three geochemists proposed an explanation[3] right out of a "B" science fiction movie. They suggest that it's beginning to look a lot like Christmas. It is snowing metal.

The first line of evidence comes from radar imaging of the planet. Images show a strange pattern of brightening on high ground, beginning at about 3½ km above the planetary "sea level." In radar, bright reflections usually mean a rough surface, but something else is going on here. The brightening blankets everything from rugged mountains to high plateaus. A variety of metals could explain the radar returns. At Venus's drastic pressures, specific metals called halides and chalcogenides may exist as vapor. On Venus, as on Earth, temperatures drop with altitude. Low-lying plains on Venus register a blistering 873°F, while the bright, higher elevations cool down to a lovely 728°F. Volatile metals may vaporize in the lowlands and migrate gradually to higher terrain, condensing again as they cool. A haze of metallic vapor could explain the mysterious 12.5 km anomaly and would certainly add to the alien nature of Venusian meteorology.

A second possibility has nothing to do with Venusian weather, but rather with the hardware design of the probes themselves. University of Wisconsin planetary researcher Larry Sromovsky authored the final report from a

3. See *Volatile transport on Venus and implications for surface geochemistry and geology* by Brackett, Fegley, and Arvidson, *Journal of Geophysical Research*, pp. 1553–1563, January 25, 1995.

NASA workshop addressing the issue. "All anomalies were attributed to an insulation failure of the external harness. As I recall, a technician decided on his own to 'improve' the external connections by using shrink tubing. This tubing could not stand the high temperatures reached during descent, decomposed, and shorted out the connections."

A hardware failure common to the Pioneer probes does not seem to solve another aspect of the phenomenon. The Soviet probes Venera 11, 12, 13, and 14 all experienced similar power spikes at about the same altitude, and the VEGA 1 Venus lander may have prematurely triggered its landing science sequence. It attempted to drill and analyze rocks some 18 km above the surface!

Another riddle for Venus aficionados to solve concerns water, past and present. Determining the water content of a planet's atmosphere is a difficult task, as we will see with Jupiter. The amount of water in Venus's air, from its formative years to modern times, is important to know because, as we have seen, it may have governed how much carbon was floating around in the Venusian system, in air and on land. The water levels also rule the skies by determining how much sulfuric acid – in hazes and clouds – drives Venusian weather. Water is an indicator of a planet's developmental history, and it affects changes in climate. Nailing down the atmospheric water was an important goal of Pioneer Venus's largest probe, called the Sounder. Orbiters had revealed a parched upper atmosphere, but how much water was below the cloud decks?

The probe carried a mass spectrometer to measure levels of various gases in the atmosphere (see Chap. 4). But Murphy's Law, that universal cosmic axiom, was at work: a droplet of liquid got sucked into the intake tube and

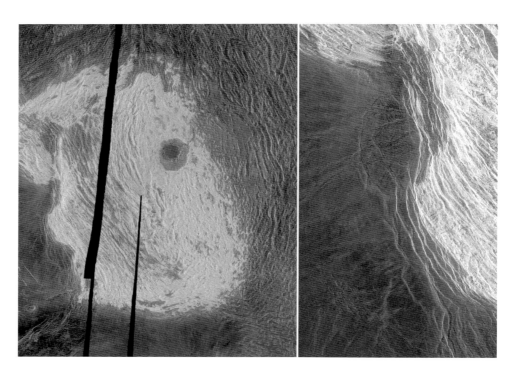

The highest mountain group on Venus, Maxwell Montes, rises 6½ miles into gloomy skies. Upper slopes of the mountain complex reflect radar in ways that imply a frost of metal. Note the transition from low to high altitude in the detail at right (NASA/JPL)

blocked it for most of the probe's descent. The instrument dutifully carried out its experiment, sampling the air, but that little drop was evaporating in the mix, and may well have thrown off its readings. The mass spectrometer's results were somewhat suspect.

Scientists tried to decipher the jumbled data for years. They concluded, with caveats, that just below the cloud deck, water is at the 100 ppm level, drier than the most drought-stricken day in the Sahara. Toward the ground, things just get worse, with water dropping to 20 ppm. The results are bewildering; because of the mixing of gases near the surface of a planet, the water levels should either be stable or increase with altitude, rather than decrease. If water is more abundant at one altitude than another, the implication is that it is flowing through the air. The decreasing numbers might work if water were being pumped into the air from the cloud deck and being removed from the air near the surface. But the numbers don't work out that way. Water levels would need to be pouring from clouds that are made up of mostly pure sulfuric acid and being sucked out of the atmosphere at prodigious rates near the ground. The mystery only deepens with results from several Soviet Venera probes. The Venera landers watched the changing spectrum of sunlight as they drifted farther and farther into the deep atmosphere. They radioed back similar results.

Although Venus may be suffering a planetary drought today, researchers naturally wonder if the planet ever had liquid water, and perhaps oceans, on its surface. The acid raindrop that clogged Pioneer's inlet also gave the mass spectrometer an opportunity to do a careful analysis of it. As the droplet evaporated, the mass spectrometer sensed its various components: hydrogen, sulfur, and oxygen make up sulfuric acid, and they were all there as they should have been. But within the hydrogen was deuterium, also known as "heavy" hydrogen. Normal hydrogen atoms, the original building blocks of the universe, have one proton and one electron. Deuterium is identical, but adds a neutron, making it twice as heavy. The ratio of deuterium to regular hydrogen changes over time, because regular hydrogen is light enough to escape into space. It leaves behind the heavy hydrogen, so the ratio of heavy to light hydrogen increases over time. In Earth's oceans, for every deuterium atom there are 6,000 hydrogen ones. The ratio is 120 times greater on Venus, meaning that Venus has a great deal of the heavy variety of hydrogen. The conclusion is that Venus has had to lose a lot of hydrogen to have so much deuterium left behind. All that hydrogen, the argument goes, must have been locked up as water in great seas or oceans.

This is one possible explanation, but as is so often true in science, it comes with its own set of problems. First, no one really knows how much deuterium and hydrogen Venus started with. The estimates of how much hydrogen has been lost rely on assumptions about conditions on early Venus. This is a bit of a jump. After all, we're not even sure about conditions on current Venus! The second problem is that the hydrogen may not have been steadily leaking away over the lifetime of the planet. Hydrogen may be fed into the atmosphere by sources that are, as yet, unknown. For now, the

Mariner 10 imaged Venus in the ultraviolet, giving us our first clear look at cloud bands within its streams of hurricane-force winds (NASA/JPL)

mystery of Venusian water remains just that: a mystery.

To the naked eye, Venus is a featureless, glistening disk. But fortunately for us, the clouds are stained by mists that are visible in ultraviolet light. The material that is absorbing UV light is somewhat of a mystery in itself. Whatever the substance is, it absorbs fully half of the solar energy falling on the face of Earth's sister. It also enables us to see the motions of the clouds, and those motions are frenetic.

Kevin Baines is a co-investigator on Europe's Venus Express mission. ESA's highly successful Venus orbiter is providing insights into the hectic nature of Venus's alien skies. "Near the equator of Venus, the clouds look very cumulus. There are big convecting clouds of sulfuric acid. As these clouds start migrating poleward they get stretched out east/west. They become thin and linear. Maybe the changes happen at night; we don't know. Venus Express can measure clouds at night in the infrared, but these clouds are at a different level from the ones in visible light. It looks like all the clouds congregate in one big mass at the poles."

That cloudy congregation at the poles swirls into massive maelstroms, vortices centered on each pole, and we have been studying the phenomenon with Venus Express, along with data from other Venus missions. "Essentially the view is that the entire hemispheric circulation of both hemispheres north and south is like a giant vortex. People can call it a polar vortex and think that its limited only to the polar region, but it's so big that the whole region is situated over the pole. If you look at it in a polar projection there is no boundary to it. The in-flow region starts just near the equator. There is no equatorial limit to it. It goes all the way to the equator. It's like a giant vortex. And because each field of sunlight emits heat from the surface, it has no lack of energy, so that's why it's in a perpetual state, like a hurricane. But dynamically a vortex is a vortex. It doesn't matter how you create it. They show the same properties. And that's something that we had shown on Venus; you see the dynamical instability that you see in a cyclone." A similar feature has been found at the south pole of Saturn, an opening in the clouds shaped like an eye. But Limaye points out a difference on Venus. "It's like looking through a haze; even though there is an eyeball cloud you wouldn't be able to see it. Dynamically you can see it through the atmospheric motions, but not in terms of the cloud. The clouds are not allowing you to see it. They don't evaporate to reveal the lower topography because even if there is the evaporation of the upper cloud particles, there are still denser parts deeper down so you can't see any detail. That's the problem. Still, you can see it in the thermal emissions."

Venus Express is able to track the clouds at different altitudes with an instrument called VIRTIS, the Visible and Infrared Thermal Imaging

Spectrometer. VIRTIS observes the planet at several wavelengths, each of which penetrates the cloud layer to a different altitude. VIRTIS has been able to chart the motions of hundreds of clouds in each layer. VIRTIS discovered that winds are stronger in the upper clouds at equatorial latitudes with wind speeds of 110 m/s. They decrease to 60–70 m/s in the lower cloud decks near the equator. Further north and south, winds are virtually the same in all three layers, decreasing to zero as they approach the pole.

The cloud structure of Venus appears to vary quite a bit from location to location. Of the five Venera probes and four Pioneer probes to survive a transit through the clouds, none gave the same results. All craft monitored the majority of the sky through all three cloud decks, and all probes recorded significantly different profiles. The clouds on Venus are always in turmoil, ever changing.

The clouds themselves are remarkably diffuse, far less dense than terrestrial ones. On Earth, average cloud droplets are similar in size to the largest Mode 3 droplets in Venusian clouds. Earthly clouds are, on average, some ten times the density of those on Venus, and can reach one hundred times the density. Rather than our familiar billowing walls of opaque white, the clouds of Venus are more like thin fog or haze stretching from one horizon to the other. The reason we cannot see the surface of Venus from the outside is not because its clouds are dense, but rather because there is so much of them. Aphrodite's foggy veil extends vertically for 13 miles.

The streams of clouds drifting on Venus's hurricane-like winds take on elegant forms. Spirals unwind over the poles. Great rivers of air spread across the equator in sideways "Y" shapes or curl across it in a giant "C" pattern. At the point where the Sun is directly overhead, the atmosphere births polygonal shapes, perhaps due to upwellings beneath the Sun's fierce heat. The geometric shapes follow the Sun's course across the planet.

Within the Venusian clouds, many scientists expected to detect lightning. Several Venera probes carried an experiment called GROZA, which indirectly detected cloud-to-cloud lightning bolts 2–32 km in altitude. Some researchers claimed that Venera 12 recorded up to 1,200 separate

The southern polar regions of Venus through the eyes of ESA's Venus Express (ESA)

Clouds seem to congregate at the poles of Venus (Venus Express, ESA)

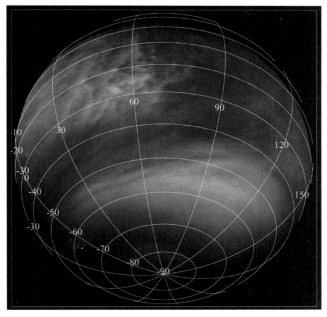

Caught in the Act?

In 1978, the Pioneer Venus orbiter may have arrived just after an earthshaking event. The Venusian atmosphere was infused with huge quantities of sulfur in the form of sulfur dioxide, but the orbiter's ultraviolet spectrometer watched as global sulfur dioxide levels steadily dropped. Over several years, sulfur dioxide continued to disappear from Venus's upper atmosphere. Why was this happening? It could not have gone on for long; if it had, sulfur levels on the planet would have had to start at unnaturally high levels. One possibility – first proposed by Pioneer Venus investigator Larry Esposito – was that Pioneer Venus arrived in the aftermath of a major volcanic eruption. Volcanoes raise sulfur levels on Earth all the time. Venus may have been showing us that much of its sulfur comes from beneath, Esposito explains. "It is still a hypothesis that has not been confirmed and more recent missions have looked for volcanoes exploding. They haven't gotten the same quality of evidence that we got from Pioneer Venus, so it's still speculation. But Venus is geologically active. The question is whether it was active while we were actually looking at it. I still think that's a likely

explanation for our results." Another possibility is that Pioneer was observing the latter stages of a natural cycle in which sulfur from below is brought to high altitude, and then sinks back down again. No one knows what the true cause of the disappearing sulfur was, and scientists are still searching for a "smoking gun" that would indicate active volcanism on Venus, Esposito says. "Some people would say it's purely an atmospheric effect – that is, the dynamics of the atmosphere stirred up the Venus atmosphere to bring sulfur dioxide into view. We would say of course that's happening, but it's the volcano itself that causes that overturn in the atmosphere. It's the volcanic eruption that creates the heat that mixes up the sulfur dioxide above the clouds where we can see it. So I'm not denying there's an atmospheric mixing phenomenon here. My hypothesis is that to move the atmosphere so much, a volcanic eruption is the best explanation. There really hasn't been any confirmation of that – no smoking gun. We're still looking for either an individual piece of evidence that would confirm that or a more general understanding of Venus, that could say something about how active the surface is. Is it really active – even as we watch?"

The Noble Gases: Where They All Came From

Outer planets researcher Kevin Baines explains the value of noble gases such as argon, krypton, and xenon. "One thing we want to do is find out where Venus came from, its origin and evolution. How did it form? What's its relationship to Earth and the other terrestrial planets? To do that, you have to find molecules that have been around forever and haven't changed. Those are called noble elements – helium, argon, krypton, neon, and xenon. Whatever their abundances are now, that's what they were 5 billion years ago. You can find out if Venus came out of the same primordial soup that Earth did by looking at some of these molecules, comparing them to Earth and Mars." Part of the challenge, says veteran planetary scientist Ben Clark, is that the measurements must be taken directly. "You cannot see argon from

orbit. You can see nitrogen and you can see CO_2, but there are no spectral lines for argon. It has to be taken in situ."

Thanks to various atmospheric probes and landers, beginning with Viking in 1976, we have fairly detailed information about noble gases on Mars, but similar data for Venus is still lacking. Researchers would especially like to ascertain the level of xenon, Baines says. "It turns out that we have a number of these [noble gas levels] from Venus, but the most telling one, xenon, has nine isotopes. You can do more detailed studies with it. Xenon is like a telltale Rosetta Stone into the history of Venus. We really want to go back and measure xenon. We also want to measure krypton. Between the Soviet Union and America, nobody really did the noble gases very well. They tried, but all the experiments failed."

strikes, but this claim has been disputed. Data from the Jupiter-bound Galileo spacecraft also indicated active lightning.

WINDS

Despite the planet's lethargic rotation, the atmosphere of Venus races across the planet at breakneck speed. Winds carry its clouds completely around the globe every 4 days. This phenomenon, known as "superrotation," was first discovered by JPL's Victor Kerzhanovich while he was at Moscow's Institute for Space Research. Kerzhanovich came up with the idea of measuring the drift of Venera spacecraft during their descent using the Doppler shift in their radio signals. It was not an easy task. "When you measure the frequency, it is a component of the velocity of the lander and the location of the observer. So one factor is wind velocity, but the other factor which is often much more significant is the relative Earth/Venus motion and the vertical velocity of the lander itself. You have to subtract all of these components out to get wind velocity. The problem with the first four Soviet Veneras was that, first, the oscillators we had in those days were very unstable, so I had to study them for many years to make them useful for measurements. The other problem was that the wind experiment was not an official part of the Soviet program in those early missions, so the first four Venera were aimed directly under the sub-Earth point. They tried to hit the center of Venus as seen from Earth. That's fine for communications, but for wind measurements this is terrible because you don't have the wind component in two directions. Fortunately, they missed several times, by about ten degrees. We probably could have discovered the superrotation earlier, but the accuracy of oscillator combined with the flight geometry was pretty bad." Venera 8's descent carried it about 40° away from the sub-Earth point, so that wind velocities could be discerned with much greater accuracy. That's when the surprise came, Kerzhanovich remembers. "All of my teachers, all the people with much bigger authority in Venus dynamics, everybody thought the Venus winds were 2–3 m/s in the bulk of the atmosphere. French astronomers had some idea of superrotation in the upper atmosphere with ultraviolet studies. But the actual superrotation happens at all the heights, even below the visible clouds, all the way down to 10 km above the surface. Winds are at 100 km/h where pressure is 16 times more than the surface of Earth. This is huge. It's like Earth's ocean with the speed and energy of 10 m/s. It is huge kinetic energy. At first we didn't trust our numbers. It was completely different than any models. Even today, up to now, the mechanism for this is unknown."

Superrotation is one of the most striking features of Venusian meteorology, says Kevin Baines. "Venus has this outstanding problem in planetary fluid dynamics, and that's the superrotation. There are lots of bizarre things about Venus, but the superrotation is one of the most bizarre. No matter where you go on Venus, from the equator to the pole, and anywhere from

100 m up from the ground to 100 km up, you would experience hurricane-force winds. This planet is rotating slower than you can walk – about 4 miles per hour. And yet, the winds are howling at 100 m/s. There's a factor of 60 between the rotation of the planet and the winds it generates."

This remarkable movement of air continues to challenge atmospheric dynamicists and planetary scientists across the globe.[4] The first truly accurate record of Venus's air currents came to us via two balloon probes, the first such craft to visit another world. The Soviet VEGA 1 and 2 missions carried French-built balloon probes (see Chap. 4). The VEGAs carried landers based on the Venera design, along with a flyby craft designed to continue on to encounters with Halley's Comet. The landers each contained a balloon probe that inflated during descent. The balloons floated at an altitude of about 54 km, inside the middle cloud deck. Beneath each 3.4-m-diameter balloon, a 1.3-m tall gondola charted temperature and pressure. The pressure remained at a fairly constant 535 mbar (remember that pressure at Earth's sea level is about 1,000 mbar, or 1 bar). Temperatures ranged from 80 to 113°F. (VEGA 2s temperatures were consistently higher, perhaps due to effects from features on the ground below.)

The balloons were specifically designed for these relatively benign conditions. A nephelometer sensed particles in the air around the gondolas. Radio systems on the tiny craft enabled ground stations on Earth to track the balloon's progress across Venusian skies. To Kevin Baines, this was the most valuable aspect of the mission. "The main thing [the VEGA balloons] did was dynamics: they had this 'trace particle' in the atmosphere. That's very important because the way we trace winds in all these atmospheres is we look at clouds. But in places like Neptune, clouds are so ephemeral that you don't know if you are measuring it right. So people would love to have their own little test particle fly around and they'd know exactly what it was. This is what you had with the [VEGA] balloon: a test particle that you could really follow using radio techniques, and you could accurately tell what the winds were really doing." Twenty tracking stations around the world monitored the balloons as currents carried each across distances of over 11,000 km (6,820 miles). Winds were fierce, howling at 66 m/s.

The horizontal motion of Venus's winds was already understood at a rudimentary level when the VEGAs arrived. But the extent of vertical currents was unknown. Both balloons hit updrafts, or vertical winds with speeds reaching 2 km/s. "Everything has been a surprise," says the French Space Agency's Jacques Blamont, inventor and chief investigator of the VEGA balloons. "Nature has proven to be different everywhere from what we expected."

Since the VEGA missions, other orbiters have refined our understanding of Venus' dynamic weather. From orbit, spacecraft can chart east/west winds fairly accurately. Just about 99% of Venus's winds flow in an east/west direction. Scarcely one percent of the air currents blow north or south. Kevin Baines, a team member of ESA's Venus Express orbiter, believes that discerning this minority flow is critical to our understanding of how the planet's energy gets from equator to the poles. "Although we've looked for 3 years

4. If there are any atmospheric dynamicists on the Venusian globe, they probably have it figured out.

now, Venus Express has not come up with any good answers about how fast the winds are blowing towards the poles. The poleward motions are down in the order of a meter or 2/s. Eastward winds are from 70 up to 100 m/s. It's very hard for Venus Express to measure that poleward motion because it's in a non-compliant orbit, a very polar orbit. If you want to look at clouds you'd better be sitting over those clouds for a long time to watch them move. The spacecraft is moving from pole to pole and can't see it."

Venus Express is also shedding light on past wind data, which seemed inconsistent. According to Baines, winds varied dramatically from probe to probe. "The data from Venus Express is giving all different results of wind speeds and directions, with fairly wide error bars. We're starting to realize that it's not measurement error; the reason we're getting all these error bars is that the planet really does have winds that are changing. They change from day to day and week to week, from latitude to latitude, gusty one day and steady the next."

Sanjay Limaye emphasizes how difficult it is to analyze the motions of the complex atmosphere. "It's difficult, if not almost impossible, to get good estimates because there are no cloud features that you can track. There's a cloud cover, but it's like fog. You see movements that are somewhat chaotic and extremely turbulent. It's like switching on lights in a room. You see a light here and then half an hour later you see a light there. The question is, did the light itself move around or are there two separate sources of light? That's really the question. And normally, when the flow is well behaved, then you sort of expect the continuity. So if something is moving in one direction you expect it to show up at a certain time. That requires almost continuous observation. That's something we haven't had yet."

GEOLOGY AND WEATHER

While spacecraft carry out direct studies of distant atmospheres, researchers gain insights about short- and long-term weather from another source: the ground. An early lesson from planetary science can be summed up in a simple statement: *The surface below informs us of the weather above.* If we see river channels and flood plains, this tells us that the atmosphere above has, at one time, had an active cycle of rainfall, evaporation, and condensation. Surfaces battered by many craters have had less atmosphere throughout their history than those with few craters. Various salts or other chemistry in the soil can constitute telltale signs of past standing water, which requires a specific set of meteorology. Glaciers, which need a stable environment with snowfall over long periods, leave specific erosional features such as moraines and "hanging" (U-shaped) valleys. Glacial valleys and runoff channels tell us that the weather on Mars is different now than it was in the past.

Venus and Mars have extensive volcanic structures created over long periods of time. Some of these structures may be dormant or active even today. Volcanoes contribute to a planet's atmosphere, so the level of past and present volcanism informs us about past and present atmospheric conditions. Canyons, mesas, patterned ice features, and sand dunes all have lessons to teach about wind and erosional rates. Titan's sand dunes provide cosmic weather vanes for its winds, and the dunes of Mars may supply clues to its climatic history.

But the relationship between geology and atmosphere on Mars runs even deeper, says Bill Hartmann. "The interpretation of the geology at small scale – the 10 m scale topography in geology is intimately tied to the climate science. I guess I could almost say you can't understand the one without the other. I think you certainly can't understand the surface topography and geology you are seeing, without understanding the climate cycles." Water is a common factor in geology, weather, and long-term climate. It is a marker of temperatures and of atmospheric behavior. As such, the behavior of water ice in the ground is a critical piece of the climate and meteorology puzzle, Hartmann contends. "You have this ice layer at the base; there could be depths where the top of that ice layer is actually melted with water circulating, and I think that is still an open question: what's going on there? Because, after all, if the ice melts and turns into liquid or water vapor, that could percolate up through the loose soil and freeze at the surface. So that could be another source of getting ice in the top 10 m even at lower latitudes as opposed to getting it there from the atmosphere. The whole issue of the climate science and the sub-surface geology science – those are intimately linked together."

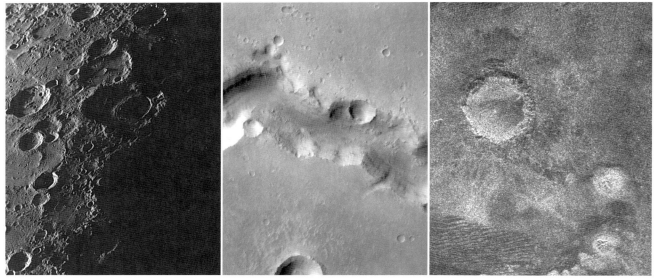

Although other surface processes are at work, cratering rates and preservation can tell us how much active weather a planet or moon has. Left to right: Airless worlds such as Mercury are saturated with fresh-looking craters (JPL); Mars weather erases some of its craters, and past weather patterns have left drainage valleys (Arizona State University); craters on Titan, whose atmosphere is dense, are rare (NASA/JPL)

As we look to alien skies, we must also look to alien landscapes to help us decipher meteorology on other worlds.

SAVING THE OTHER PLANET

When the Pioneer Venus orbiter began circling Venus in 1978, it discovered something unsettling. Enormous holes punctured the planet's ozone layer, one over each pole. On Earth, ozone stands as an atmospheric shield, staving off the deadly effects of the Sun's ultraviolet radiation. Venus should have had the same kind of high-altitude shelter, but something had torn openings in the planet's ozone umbrella.

Our probes told us that the complex chemistry of Venus's environment naturally creates chlorofluorocarbons, known as CFCs. CFCs are compounds containing chlorine and fluorine. We use them in our refrigerators and air conditioners. In this form, they are inert, and at one time they were thought to be safe for the environment. But on Venus, Pioneer spotted CFCs in action. As sunlight fell upon those floating compounds, it tore them apart into chlorine and fluorine. These two gases, acting by themselves, are not inert. In fact, Pioneer observed them breaking down Venus's ozone layer. Coincidentally, manufacturers on Earth had just begun widespread use of CFCs, sifting it into hair sprays, deodorants, pesticides, and paint cans and other pressurized products. Without knowing it, Earth sat on a nice pile of self-made dynamite, and we'd just lit its chlorofluorocarbon fuse.

Several researchers got the brilliant idea that if Venusian CFCs were tearing away the ozone layer on Venus, they just might do the same on Earth. We now know this to be true. Some environmental chemists had tumbled to the process earlier, but their warnings had fallen on a deaf population. The world that embodied the tempestuous goddess of beauty gave us a stark example of what could happen to our own fragile world. Because of the study of Venus, industry did an about-face, saving our atmosphere from thousands of tons of ozone-grabbing chemicals. Pioneer Venus underlined the fact that the study of planets, in a very real sense, is the study of Earth. Applied wisely, the knowledge we gain by scrutinizing distant worlds can help humankind understand and take better care of this one.

Chapter 6

Mars

The Noachian epoch saw a nightmarish hail of rock and metal from the sky and volcanism from below (© Michael Carroll)

With the end of autumn, the Sun has bobbed along the horizon for many weeks. Sunsets on Mars are blue, and these winter days see more and more blue skies. Nights are growing longer than days, and new weather encroaches – it is beginning to snow. Dry-ice frost forms on hexagonal ridges stretched over nearby ground, telltale signs of subsurface permafrost. Ice crystals grow on metallic landing legs and bloom over mirrored solar panels, sprouting

M. Carroll, *Drifting on Alien Winds: Exploring the Skies and Weather of Other Worlds,*
DOI 10.1007/978-1-4419-6917-0_6, © Springer Science+Business Media, LLC 2011

like crystalline ferns. Soon, the Phoenix lander will be encased in a coffin of carbon dioxide ice. Winter has come to northern Mars. But Martian winters have not always been this way.

The story of Mars's atmosphere is one of remarkable transformation. To understand the Martian meteorology of today, we must travel back in time to a Mars 3 billion years distant. That ancient Mars has left us clues to its nature. With Mariner 9's first global reconnaissance of the Red Planet (see Chap. 3), cratered terrain surrendered before vast flood plains, dendritic riverbeds, and spreading deltas of fossilized waterways. Many of these drainage features formed early in Martian history, in the first of three geologic epochs. The Noachian epoch marked the formation of the most ancient Martian surfaces visible today. It was a nightmarish time of molten rock, downpours of stone and metal asteroids, and violent volcanoes. Whatever primitive atmosphere Mars harbored, left over from the Solar System formation, was either stripped away by massive impacts or leaked away into space under Mars's low gravity, just a third that of Earth's.

A new atmosphere arose, belched from the interior through the throats of volcanoes. Toward the end of this period, the Tharsis bulge, the largest volcanic construct of any planet, rose from the cratered Martian plains. This planetary beer belly would play an important role in future meteorology. Late in the Noachian, waters roared across Martian lowlands in extensive flash flooding, engraving the plains with a calligraphy of riverbeds and canyons. Seas may have circumscribed the Martian north; remnants of rivers seemed to flow into the area from the southern highlands. If this was the case, the weather was dramatically different, according to Lockheed/Martin's Ben Clark, whose involvement in Mars spans a period beginning with the Viking landers. "Part of the mystery is: 'Could the Martian atmosphere have been thicker at one time?' The current thinking is that it was, but that once Mars lost its magnetic field, that allowed the solar wind to erode away a major fraction of the atmosphere." The issue that researchers like Clark are puzzling over is that the carbon dioxide should have been locked into the rocks by now. "It's a little bit of a dilemma in the sense that we have CO_2 in the atmosphere, and we know it can react with the rocks to make carbonates. Why, then, didn't the carbon dioxide disappear altogether? On Earth, we have biology recycling the CO_2 for us. The plants take it up and the animals eat the plants and convert it right back to CO_2. It just goes in a circle. On Mars you don't have that."

Volcanoes are responsible for CO_2 in the atmospheres of Earth and Venus, and Mars does exhibit a record of volcanism, with the largest volcanoes in the Solar System rising high into the Martian sky. In the Tharsis region, Olympus Mons is more than twice as high as Mount Everest, with three nearby volcanoes in close competition for altitude records. On the opposite side of the globe lies another volcanic region, the Elysium province. Other volcanic features in the south are more ancient, with heavily eroded mountains, weathered lava flows, and fields of cinder cones. The volcanism on Mars appears to be extensive, but as Clark points out, "It's

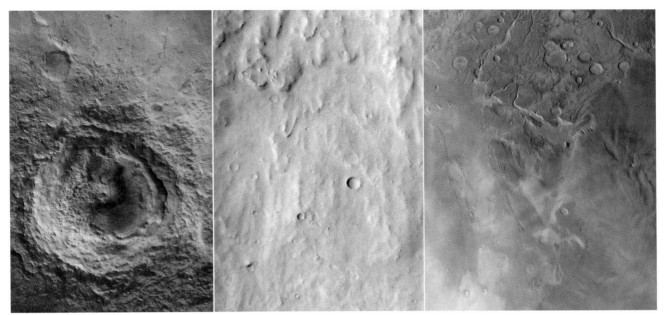

Three views of Martian terrain of different ages. Left to right: Maunder Crater, Noachian; Hesperian lava flows; and Amazonian-age features in Amazonis Planitia (ESA/DLR/FU Berlin, G. Neukum; NASA/JPL/ASU; NASA/JPL Viking 1 photo)

really a lot less than we've had on Earth, at least ten times less and perhaps a hundred times less. Still, if you heat rocks hot enough in volcanoes the CO_2 will come back out of them. That could be part of the answer."

If volcanism explains the presence of Martian CO_2, the planet's atmosphere may have been pumped up substantially in the next epoch, the Hesperian. About 3.5–1.8 billion years ago, extensive lava plains spread across the face of the Red Planet. During this period, Mars was a volcanic overachiever. Eruptions during this period likely built the massive volcanoes sitting atop the Tharsis province. The great bombardment of meteors, comets, and asteroids tailed off at the beginning of the Hesperian, so that most craters from this era are heavily eroded.

Beginning about 1.8 billion years ago, the Amazonian epoch continued to be marked by lava flows. Of the rare impact craters that survive from this quiet period on its relatively recent surfaces, many are well preserved compared to those of the Hesperian or Noachian. But volcanoes eventually died out. No active volcanism has been observed on Mars in modern times, although the search continues. The Amazonian epoch stretches into the present. The environment of modern Mars is, in many ways, the closest to Earth of any other world. Compared to Venus or the outer planets, Martian temperatures are welcoming, ranging – near the equator – from 128°F below zero at night to a high reaching the melting point of water. Mars's length of day is nearly identical to Earth's (24 h, 39 min),[1] and its axial tilt, which governs seasons, is also currently quite similar, coming in at 25.2° (compared to Earth's 23.5°).

Rain has not fallen on the Red Planet for perhaps billions of years. Today, frosts of carbon dioxide dust the ruddy rocks. But during periods of volcanic activity, and perhaps in between, standing bodies of water were probably

1. This is the length of its solar day, the time it takes for the Sun to return to the same apparent spot in the sky. The sidereal day, the time it takes for the planet to make one rotation compared to the background stars, is 24 h, 37 min, 22 s.

Volcanoes in different stages of erosion. Left to right: Ceraunius, with moderate cratering (Viking image JPL); well preserved vent in the Tempe region imaged by Mars Global Surveyor (JPL/Malin Space Science Systems); false-color image of Apollonius, a heavily eroded volcano (ESA/DLR/FU/Berlin, Neukum)

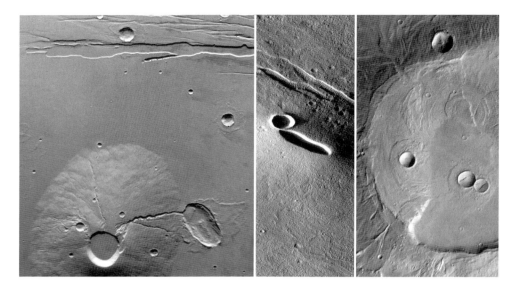

stable enough to trigger an active water cycle in the distant past. Some of those bodies of liquid may have been large enough to qualify as oceans. Flat lowlands encircle the northern polar region of Mars. Few craters exist in this, the most geologically young region of the planet. There is only one other place in the Solar System where we see the same kind of flat profile in topography: Earth's ocean basins. The terrain to the south of these lowlands is more ancient, rugged, and heavily cratered in places.

Parallel ridges may mark an ancient shoreline of a great northern ocean (JPL)

The two-faced nature of the Red Planet is called planetary dichotomy. At the boundary between ancient cratered highlands and smooth northern plains, fossilized river valleys drain toward the lowlands in a web of branching valleys, deltas, and oxbows. Bench-like features similar to Earthly beachfront property rim the plains. Many researchers feel these features confirm a time when vast oceans ringed the north pole, with rainstorms feeding rivers, lakes, and a great expanse called Oceanus Borealis (the candidate northern ocean). At least two possible major shorelines of differing ages have been identified, each a thousand miles long, roughly the distance from Seattle to Los Angeles.

It is clear that Martian meteorology has changed dramatically over the lifetime of the planet. No liquid water can remain on the Martian surface today; it instantly boils into vapor in the thin air. But past epochs may have seen enough pressure and water vapor to result in precipitation. Some river valleys are clearly the product of flash floods that may have been triggered by the melting of subsurface ice from volcanic eruptions, subsurface ice melt, or the impacts of meteoroids. But others open out in gentle fans, implying long-term precipitation. Some valleys appear to be glaciated. Glaciers also require stable climates with long periods of snowfall. Debate continues to rage over just how much water was there in the past, how long that water lasted, and what happened to it.

Some Martian valleys adjacent to the Argyre impact basin appear to have been carved by glaciers, indicating a longstanding water cycle (© Michael Carroll)

Orbital studies by Mars Global Surveyor, first reported by Malin Space Science Systems, have uncovered evidence of active surface water even today. Gullies on crater walls changed from one imaging session to another, showing shifts in drainage patterns and brightening of surfaces, implying that ice has condensed as groundwater escapes from the crater's subsurface. A gully on the wall of an unnamed crater in Terra Sirenum, at 36.6°S, 161.8°W, was initially imaged by the spacecraft in December 2001. It showed a group of gullies dusted by light-toned material. A second image was radioed to Earth in April 2005 and again in August, confirming that new bright material had been deposited. The MGS images show that a material flowed down through a gully channel, then spread out near the base of the flow.

CAPPING IT ALL OFF

Martian weather is governed by the same factors that Earth's is: temperature gradients; solar heating; atmospheric hazes, clouds, and flow; greenhouse warming; axial tilt; and length of day. But Mars has yet another important

Gases from lava flows and their volcanic sources must have contributed substantially to Martian air pressure in the past (NASA)

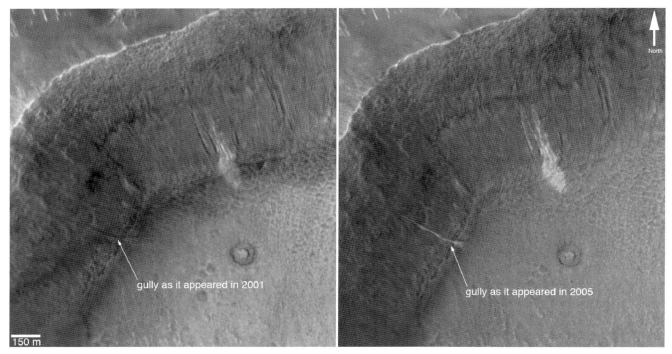

150 m

gully as it appeared in 2001

gully as it appeared in 2005

North

Changes in crater walls may indicate active water flows today (MSSS/JPL)

An artist's reconstruction of the event captured in the previous images (© Michael Carroll)

meteorological influence, a surface feature intimately linked to its weather and climate – its polar caps. Mars is cold enough for carbon dioxide to freeze onto the surface as dry ice. Its polar caps are a mix of water ice (concrete-hard in the Martian environment) and carbon dioxide ice. In winter, carbon dioxide plays a similar role on Mars to that played on Earth by water. In the terrestrial air, water vapor condenses into snow or frost that makes its way groundward, covering the surface. On Mars it is the carbon dioxide that condenses, snowing primarily on the polar regions.

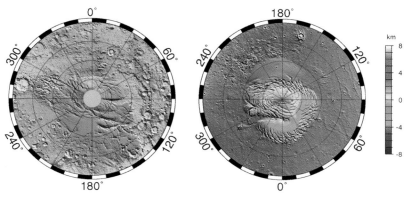

Topographic map of the south (left) and north polar caps from laser altimeter data taken by Mars Global Surveyor (USGS)

The atmosphere on Mars is thin, roughly equivalent to the pressure on Earth at 100,000 ft. Carbon dioxide in the polar caps constitutes a substantial portion of this atmosphere, says California Institute of Technology planetary scientist Andrew Ingersoll. "On Mars, there's an annual deposit of CO_2 that's put down in the winter time. When the Sun hits it, it evaporates and goes into the atmosphere. At any one time, there is more carbon dioxide in the atmosphere than there is in this annual cycle of carbon dioxide. The atmospheric pressure goes up and down on Mars by about 30%."

The northern and southern caps are not identical. Both have a core of frozen water coated by carbon dioxide ice, but seasonal differences affect their structure. Because of the eccentricity of the Martian orbit, southern winters are longer and colder than northern ones. This leads to a layer of CO_2 ice on top of the residual water ice cap in the south. There is a thicker seasonal ice cap of CO_2 there as well. Some of the carbon dioxide and water freeze to the surface of the cap on contact, or condense directly from the air. But is it possible that some precipitates out, falling as snow?

Ever since Mariner 9 first hinted at a Mars with a wet past and active present, experts have debated the existence of snowfall on Mars. The Mars Global Surveyor has put the debate to rest. Its radar system imaged the clouds at high latitudes on the night side of Mars. The radar signature that reflected back indicated solid particles in the clouds. When morning came, the clouds dissipated. Combined with infrared and temperature data, along with sophisticated computer modeling, analysts believe MGS has conclusively found snowfall – made of water ice – falling from the skies of Mars.

When the Phoenix lander touched down in the northern autumn, scientists hoped to see some precipitation at the onset of winter. The solar-powered lander was not equipped to withstand the long periods of darkness or low temperatures of the upcoming winter nights. Would the lander survive long enough to witness snowfall on another world? Flight engineers knew it would be a race against time, says Phoenix Principal Investigator Peter Smith. "It was really starting to snow in earnest up above as we lost our mission. We needed another month or two to really see the effects of winter coming on. We actually saw the snow coming all the way to the surface in our LIDAR." The LIDAR (Light Detection and Ranging) experiment used a laser to chart particles in the air. Data indicated that the precipitation would not be enough to accumulate into big snowdrifts. The experiment showed that only a sprinkling of ice crystals were making it down to the surface at the end of mission. Smith's team was surprised by the similarity of Martian ice crystal clouds to ice clouds on Earth. "On Earth they're at higher altitudes above the surface where the pressure's low. They grow under about the same conditions. They get crystals 50–100 mm in size, and that's the size we were seeing on Mars. The ice crystals are very similar. You wonder if there are snowflakes. We've looked for halos or sundogs, but we didn't see any. I would bet they are there, somewhere on Mars."

How Martian Meteorites Give Us a Whiff of Mars Air

Rock hounds have discovered tens of thousands of meteorites across the world. Of these, nearly 80 hold the distinction of being pieces of the planet Mars.[2] How do we know?

The tale centers around three meteorites. It begins in the French village of Chassigny in 1815. The stillness of an autumn morning was broken by "a loud report which could be likened to discharges of numerous muskets."[3] The sound – undoubtedly a sonic boom from the incoming meteor – was heard in several surrounding villages. Witnesses described a strange gray cloud to the northeast. A man working in a vineyard watched "an opaque body from which thick smoke emanated fall to the ground." The man found stony fragments strewn across a half-meter-across crater in the plowed field. Villagers collected a total of roughly 8 lb of material from the fall. A similar event occurred 50 years later in Shergotty, a small town in the Bihar province of northeastern India. On August 25, 1865, people heard sonic booms before witnessing an 11-lb stone fall from the sky. On the morning of June 28, 1911, the villagers in the Egyptian town of El-Nakhla had a different experience. A meteor apparently exploded somewhere overhead, peppering the village with some forty stony fragments all over the area. Some stones buried themselves up to a meter in the ground. The total weight of the Nakhla space rocks added up to 22 lb.

Researchers believe that most meteorites come from asteroids. Asteroids are ancient, and many appear to have formed under cooler conditions than planets. But the three meteors under study didn't fit those constraints. They were relatively young and made of rock associated with planetary formation (igneous or basaltic rock). The Shergotty, Nakhla, and Chassigny meteorites had something more significant in common: trapped gases inside matching the Martian atmosphere. When the Vikings sent back atmospheric composition data in 1976, that data included ratios of nitrogen isotopes (isotopes are gas atoms that have the same number of protons but a different number of neutrons). By comparing the stable to isotopic compositions, researchers were able to establish that the gases in these meteorites perfectly matched the ratios of constituents in the Martian atmosphere. No other place in the Solar System is known to have this specific atmospheric signature. Other elements in the meteorite gases also matched the in situ Mars data from the Vikings. The Mars meteors are literally in a class of their own, called SNC's (pronounced "snicks") for Shergotty, Nakhla, Chassigny.

2. As of 2008, the count was up to 77 Mars meteorites worldwide.

3. Chassigny quotes are from the Paris Science journal *Annales de Chimie et de Physique*, 1818.

CLOUDS

Back in the days when Mars was etched by canals and ruled by malevolent six-armed warriors, early astronomers knew of the existence of clouds on Mars. Clouds forming on the leeward side of the volcano Olympus Mons were so stable that observers mistook them for snow, branding the area with its first name, Nix Olympica, or "snows of Olympus." Other areas brightened and cleared from one day to the next. The nature of these localized condensations were only guessed at until the Space Age. Early U.S. Mariners did not see clouds during their brief flybys, although they did image haze layers above the surface of the planet. The first orbiters did see discrete cloud formations (see Chap. 3). Landers and the Hubble Space Telescope added to our understanding of Martian skies.

Mars has none of the billowing cumulus clouds of Earth, nor the abyssal hazes of Venus, but it does have a variety of cloud forms. Cosmic meteorologists have identified four general types: ground fog, convective clouds, orographic clouds, and wave clouds. Some of the most spectacular images from the Mariner 9 mission came from early morning images of canyons. The maze-like canyons of Noctis Labyrinthus displayed mysterious fog filling deep chasms. The fog likely comes from vaporizing ground frost. "It's typically in the morning, as the Sun comes up, that you get a lot of clouds," says MRO scientist Steve Lee. "It's cooler and things condense at night, so it takes a while for those morning clouds to go away."

Catabic (downward flowing) winds trigger what meteorologists call bore-wave clouds. Other wave clouds condense on the leeward side of mountains and raised crater rims. Occasionally, cirrus clouds drift across Martian plains in winter. These will seem familiar to any future Earth travelers.

Mars has another cloud system that was unknown until long-term observations began to accrue – an equatorial belt system. In earlier missions, Lee says, "it was like you were looking at little postage stamps spread across the surface. You didn't get the context of what they were doing. You didn't realize that it was this equatorial belt of clouds that formed during the aphelion (farthest point of the orbit around the Sun) period of the year. Prior to Hubble, observations from the ground with microwave instruments showed temperature changes that were fairly repeatable from year to year, and there

An assortment of Martian clouds. Left to right: Clouds drape the flanks of the volcano Arsia Mons, with the shadow of the moon Phobos at upper center (MGS image MSSS/NASA); early morning fog fills a portion of the Valles Marineris (ESA/DLR/FU/Berlin, Neukum); wispy cirrus clouds drift above the edge of Endurance Crater (Opportunity rover images NASA/JPL)

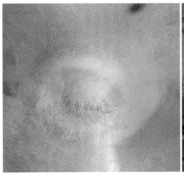 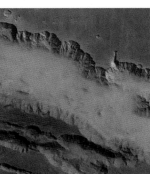  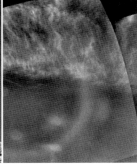

was some hypothesizing about what was going on because if it's a really cold atmosphere it means you don't have a lot of dust in it because the dust warms the atmosphere up by absorbing the sunlight. Once we got the Hubble observations, these belts of clouds just jumped right out and the temperature changes sort of all made sense."

STORMS AND DEVILS

Because of the thin atmosphere, Mars temperatures vary within inches of the surface. The temperature gradient between warm surface and chilled air sets up eddies and currents in the atmosphere near the surface. The Viking landers were equipped with sensitive seismometers to search out Marsquakes. Viking 1's instrument failed to operate. Viking 2's ran into an unexpected problem. Winds were so gusty that the seismometer essentially became a wind detector, too overpowered by alien breezes to sense movement in the ground. Viking's meteorology boom detected fast winds, though the thin Martian air had little effect on the landers themselves.

Two decades later, the Pathfinder lander charted winds in greater detail. Winds on Mars have been clocked at over 125 mph, although winds at both Viking sites and the Pathfinder site were considerably gentler. Pathfinder experienced some blustery days, says Smith. "We saw what seemed like a very large number of pressure drops coming over the lander itself. These were certainly turbulent eddies, and whether they had dust in them or not we didn't know; we weren't looking all the time. There must be a lot of turbulent eddies on Mars because of the number of pressure drops. These are little spikes [in pressure] that go down for 10 or 20 s and then come back up again. The number was just phenomenal, probably in the hundreds."

Some of the pressure drops mimicked the kind of dynamics seen in weak tornadoes or dust devils. Mariner 9 and the Viking orbiters had spotted suspicious ground-hugging "clouds" from orbit. In other regions,

Dust devils seen from orbit (l) and from the Spirit rover in the Columbia Hills, Gusev Crater (JPL)

strange tracks wound across the plains. The phenomena seemed to be circumstantial evidence for dust devils on the Red Planet. But to Peter Smith's delight, Pathfinder caught them in action. "Dust devils had been seen in the past, certainly from Viking and perhaps from Mariner 9, but they hadn't been seen from the surface, so there wasn't much known about their frequency or their height. We were the first to see them from the surface."

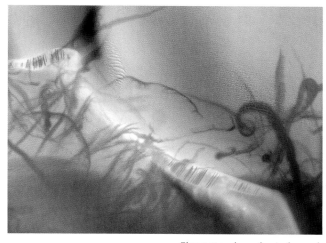

Elegant curls and spirals mark cleared pathways in the wake of dust devils, seen in this false-color image from Mars Global Surveyor (MSSS/JPL)

The Mars Exploration Rovers have had a successful dust-devil-hunting campaign. High in the Columbia Hills, the Spirit rover took chains of images every 5–10 s in an attempt to create dust devil movies. Many attempts were successful, sometimes capturing multiple mini-twisters kicking up Martian dust. Most disturbances spanned a few tens of meters wide, and wandered lazily across the plains below the rover's perch. But a few grew to colossal size, whirling in a maelstrom more than a hundred meters across.

The problem with dust devils is that they throw dust everywhere. The dust has another effect – it paints the Martian skies tan during the day and tints the sunsets a ghostly indigo/blue at dusk. Like the reddish daytime sky, the blue color at sunset is caused by the Martian dust itself. The dust in the atmosphere absorbs blue light, coloring the sky its red hue, but it also scatters some of the blue light back toward the viewer. The faint blue color only becomes apparent near sunrise and sunset, when the light has to pass through the largest amount of dust.

HURRICANE SEASON, MARTIAN STYLE

Earth has its hurricane season; Mars has its seasonal dust storms. These atmospheric disturbances dwarf those on Earth, cocooning the entire planet in a choking twilight for months at a time. The ferocity of Martian dust storms comes from two sources. First, Mars is a desert world. With millions of square miles of dunes and dusty plains, the planet affords plenty of raw material for such storms, unencumbered by such terrestrial features as ground cover, forests, and oceans. Second, the dust raises air temperatures substantially. On Earth, temperatures are controlled – in part – by water vapor in the atmosphere. The Martian atmosphere is dry, so when dust billows into the air, it becomes the dominant force, raising temperatures by as much as 30°C. The temperature change feeds more energy into the winds, and the cycle escalates.

The heat of the Sun is the ultimate power source for storms on Mars, as it is on Earth. Dust storms are ubiquitous planet-wide, but the global ones tend to germinate in the lowlands, says MRO's Steve Lee. "Over the years it seems like a lot of them occur on the lowest spots on the planet like Hellas. You can understand that because the atmospheric pressure would be higher; therefore the wind velocity needed to move things would be lower because with a denser atmosphere you're able to transfer more momentum to the surface at a given velocity. And once they start, it seems as if they can grow large enough, that they can cross the equator whether they're starting in the northern hemisphere or the southern hemisphere. Once they can cross the equator then they tend to explode and can reach truly global proportions pretty quickly." The really big ones brew just a few months after perihelion, the point in Mars's orbit when the planet is closest to the Sun.

Though more study needs to be done, many researchers believe the trigger is airborne dust particles absorbing sunlight and warming the Martian atmosphere in their immediate area. Warm pockets of air flow toward colder regions at the poles, generating winds. The winds carry more dust into the air, heating the atmosphere even more. The process forms a positive feedback loop, forcing local dust clouds to grow into planet-sized tempests of raging wind and powder. Steve Lee has been struggling to comprehend the dusty disturbances, both local and global. "One of the big puzzles comes when we look at how easy it is to entrain small particles in a wind tunnel on Earth. There's a Mars tunnel at Ames Research Center that you can pump down to the atmospheric pressure of Mars. And for the most part those results indicate that you need much faster winds than we've ever seen on Mars before anything is picked up. And yet we see dust storms happening and you see dust all over the place. You see dust being picked up off the surface. So there's obviously things we don't understand or we can't replicate with the capabilities that we have available on Earth."

Global dust storms distributed a powdery haze over the entire planet in 2001 (NASA/JPL/Malin Space Science Systems)

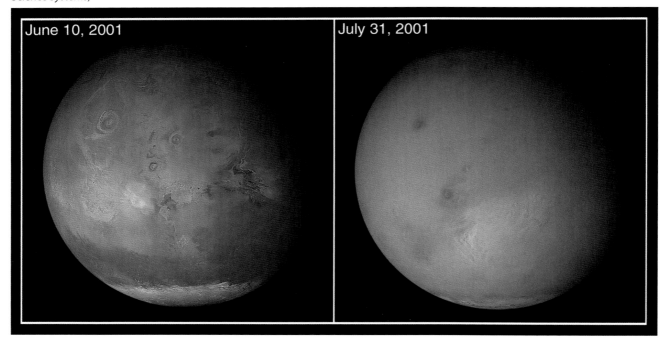

Lee and other researchers suspect that winds gather energy as they move down slope. Evidence comes from changes in surface albedo, or darkness. As a graduate student, Lee was tasked with studying the dark "collars" around the summits of the Tharsis volcanoes. These clear surfaces are darker than surrounding dust-covered areas. "They change with time, and when you look at them frequently enough you can actually catch, occasionally, little things like dust streamers blown downwind of craters, or downwind of hills or bumps in the volcanoes. So it made sense that what's causing the dark collars to form is that the wind is blowing downhill, and it eventually gets to the critical velocity needed to remove dust from the surface." Models indicate that down slope winds pick up substantial speed, enabling them to transfer enough momentum to the surface to pick up dust, even in the thin Martian air. The phenomenon happens mostly at night, when temperatures cool, Lee suggests. "What it amounts to is that at night the atmosphere cools off at the top of the volcano and gets heavier and denser. It just starts running downhill because it's denser so it wants to sink. So you're actually forming those features at night. And, moving that forward to the current day, we've been able to map those with the Mars Reconnaissance Orbiter, the MARCI instrument. And it can sort of follow them now very routinely on a couple week time-scale. You can see how they're changing almost continuously. When you have a dust storm you have increased settling of dust onto those surfaces, and they tend to retreat up the slope and almost erase themselves. And then once the dust storms are gone they start forming again very rapidly."

CLIMATE: A CHANGE IN THE AIR

The Mars of today is a desiccated desert world of dune, sand, and rock. But its climate has wild swings over geologically short time periods. To

Modern spacecraft such as the Mars Reconnaissance Orbiter are returning data in unprecedented detail, such as this view of Phobos above haze layers in the Martian atmosphere (NASA/JPL)

understand its remarkable climatic shifts, we can look to several complex factors. One such factor is Mars's elliptical orbit. Another is the lack of a large Martian moon. A third factor in a shifting climate is that bulging set of volcanoes in Tharsis. As Principle Investigator for the Phoenix lander, Peter Smith has a deep interest in Mars weather and climate change. "The climate apparently was a lot different even 5 million years ago. You can have spikes of climate change. This is all because the obliquity [the planet's tilt] is not stable." Mars spins like a gyrating top. Its axis of spin does not remain vertical, but fluctuates back and forth over periods of millions of years, Smith explains. "Right now, it's at 25.2° tilt, but not too long ago, in terms of a 100,000 years, it's been a 30 or 35° tilt, and if you go back 5 million years it might have been a 45 or 50° tilt. That means that you're in a very different climate regime."

Astronomer Bill Hartmann adds that the concept had a revolutionary effect on the planetary science community. "Along about 1999 or 2001, these French guys were showing up at the meetings and repeating a phrase that I particularly remember: 'The Mars that we see today is not the normal Mars.' This was a very profound thing because everybody from amateur astronomers in the 1950s up through 1997 when Mars Global Surveyor set off, conceptualized Mars as a static problem. We're learning about Mars. We're learning about the way Mars is. Now we've got photographs of Mars. Now we've got these river channels, etc. But it was always pictured as *this is the way Mars is*. And what I think we are still coming to grips with – and it's an underlying theme in all the talks given at all the current conferences – is that the climate is constantly changing on Mars."

Earth's axis also oscillates, but only by ±1.5°, on a cycle of about 41,000 years. Scientists have linked this subtle cycle to Earth's ice ages. Mars's extreme axial bobbing leads to diverse seasonal conditions and long-term changes in climate. Its axial oscillations take place over periods of 100,000 years, but each leads to a pinnacle of tilt every million years or so.

The forcing of climate change by the tilting of a planet's axis is called the Milankovitch[4] effect. The oval shape of Mars's orbit exacerbates the phenomenon. On Earth, seasons in the north and south are similar in intensity because our planet's orbit is fairly circular. A perfect circle has an eccentricity of zero. Mars's path around the Sun is less circular; it has a greater eccentricity than that of Earth. Earth's eccentricity is approximately 0.02, while Mars's is 0.09.

The wild tilting of the Martian axis has another cause: Mars has no large moon. Earth's Moon is nearly a quarter the diameter of its parent planet. The Earth/Moon system has been compared to a double planet. Our massive Moon actually helps to stabilize Earth's axial tilt. It takes much more energy to disrupt the tilt of two planets than one. Mars is on its own,[5] and pays for the lack of a large natural satellite by its changing climate.

The Tharsis bulge has a marked effect on the Martian spin. Like Earth's Moon (but to a lesser extent), it acts to dampen the wobble of Mars's axis. In the early Noachian, before volcanoes built Tharsis, the axis of Mars may have tilted as much as 50°, periodically forcing one pole to point at the Sun. This would burn off all the ice on that pole, releasing water vapor and any other

4. After the Serbian mathematician/geophysicist Milutin Milankovitch, who first came up with the idea.

5. Its two diminutive moons, Phobos and Deimos, are too small to have an effect on Mars's spin.

trapped volatiles. Some of these volatiles would migrate to the opposite pole, then in permanent night, where they would condense, but the rest would warm the environment and thicken the air. At other times, the Martian axis would have no tilt at all, so both of the poles would freeze any water vapor that wandered their way. These extremes would have come and gone in cycles of only a few million years, says Phoenix lander scientist Peter Smith. "The polar caps become unstable; they go away. They sublimate at a great rate, and it takes a while for them to lose 3 km of water, as you might guess. During that time there's a lot of water being pumped into the atmosphere. The frost point is going to come up – it's going to become much warmer – and you can expect a very different situation with the atmosphere and the soil. Plus, in the summers in the polar regions, with that sort of tilt, the Sun is much higher in the sky. You're going to have very warm temperatures in the summers and lots more water in the atmosphere, maybe a thousand times more."

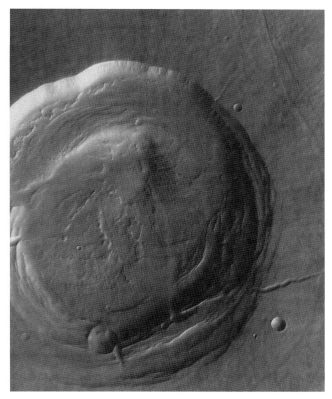

Volcanoes such as Biblis Patera in the Tharsis region had a profound affect on Martian weather and climate (ESA/DLR/FU/Berlin, Neukum)

When Tharsis rose from the Martian plains, it tamed Mars's wild ways. The off-center mass dampened the axial wobble, and the volcanoes themselves transformed the air. As on Earth, Mars volcanoes contributed water and CO_2 to the air. Estimates suggest that it took 186 million cubic miles of lava to create the entire Tharsis province. The total gas pumped into the Martian atmosphere could have amounted to a pressure 1.5 times as great as Earth currently has. In reality, the atmosphere may never have grown to such a heavy blanket of air; atmosphere is constantly leaking away from any planet. The less gravity a planet has, the more atmosphere it can lose. Mars has roughly a third the gravity that Earth does. Still, most analysts believe the air pressure in the midst of all that Noachian volcanism was plenty high enough to sustain liquid water on the surface. Ponds and lakes may have lingered even into the Hesperian. As we saw on Venus, CO_2 is a greenhouse gas. The more of it a planet has, the more heat it can retain. On Mars (unlike on Earth) this was good news. The warmth brought a water cycle and warmer weather to the little world.

And speaking of water, those volcanoes could have pumped enough water into Mars's environment to fill the northern basins with oceans more than a hundred meters deep. Tharsis and other volcanic provinces poured water vapor and greenhouse gases into the environment. As temperatures rose, ices melted. Some of those ices were dry ice (frozen CO_2) that sublimated, meaning that they turned into gas directly from ice. Others were water-ice, which turned into water vapor. Water vapor also has a warming effect. Mars was enjoying a true warming cycle.

Magnetic fields locked within Martian rock suggest that plate tectonics may have operated on early Mars (NASA/JPL/USGS)

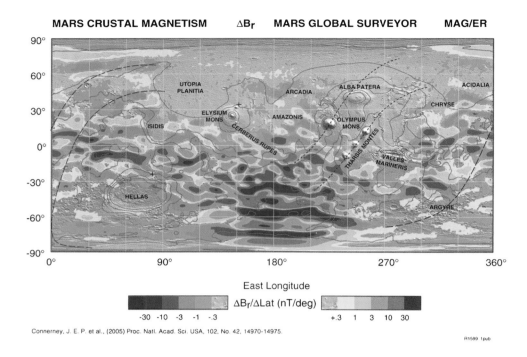

Mars may have been warmer, and may have had an active water cycle at least intermittently, into the Hesperian epoch. But all good things must come to an end. The beginning of the end for the Red Planet resulted from a critical difference between Earth and Mars: Mars has no plate tectonics anymore.

The magnetometer on Mars Global Surveyor mapped magnetic patterns locked into Mars's surface. These patterns bear some resemblance to those on the Atlantic Ocean seafloor, where two plates are spreading apart. The pattern of magnetism in Earth's ocean crust is a record of changes in our planet's magnetic field. These changes are mirrored in both plates. Similar areas have been documented on Mars, implying – but not proving – that the planet had plate tectonics at one time. When its theorized plate tectonics

Without the active plate tectonics that Earth has, Mars was doomed to lose its atmosphere (© Michael Carroll)

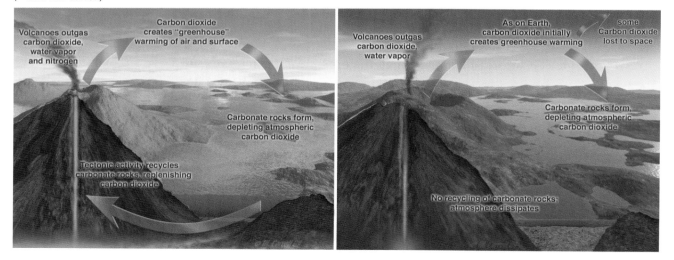

shut down is unknown. One thing appears certain – they are inactive today.

Earth's conveyor-belt-like shifting of crustal pieces melts the rock that has soaked up atmospheric chemicals, including carbon compounds. When those plates subduct (slide under each other), the gases are freed and later expelled through volcanoes and other sources. Early Mars had plenty of volcanism to support its atmosphere, but once the core lost its initial heat, there was no energy generated from tectonics to power volcanism. With less volcanic activity, much of the atmosphere was locked into the rocks or lost to space. Atmospheric loss increased as the Martian magnetic fields collapsed, because a magnetosphere provides a protective bubble around a planet. Without it, solar wind is able to impinge more directly on an atmosphere, stripping away lighter gases over time.

Meridiani Sinus, the area under study by the Opportunity rover, may have once been a vast saltpan spotted by salty lakes and ponds (© Michael Carroll)

The transition between a warm, wet Mars and the planet we see today may best be represented by the landing site of the Opportunity rover in Meridiani Sinus. After spending its first 8 weeks in the small Eagle Crater where it landed, Opportunity ventured east to the stadium-sized Endurance Crater. Studies there revealed a layered history in stone going much farther back in time. For most of the Noachian, Meridiani may have had episodes of standing water in ponds, lakes, and even flowing rivers. Upper layers of Endurance are riddled with salty minerals similar to those seen in terrestrial saltpans and dry lakebeds. But as the rover explored deeper layers, the sulfate levels started to decline. Ben Clark, a rover science team member, comments, "There's quite a gradation. One of two things may have happened: first, the salts came from above, so that there are more in the upper layers. Second, ground water may have wicked up and drew salts with it."

If the Opportunity site did constitute beachfront property, it did so long ago. Meridiani is thought to be very ancient, lying in the earliest preserved, or Noachian, terrain. Says Clark, "It couldn't have rained since crater formation at Meridiani, or you'd have washed out the sulfates deposited below them." Ponded surface water, percolating up from below, may have existed long after the Martian rains ceased.

Clark cautions that it is possible that the chemistry in Martian soil has nothing to do with past epochs of rainfall. "Some scientists tried to interpret the high sulfates and chlorides in the soil as evidence that liquid water had weathered the rock to produce salts, and then concentrated them in the surface. This is typically what happens on Earth. However, most of us believe the origin of sulfates and chlorides is volcanic gases containing sulfuric acid,

The Phoenix lander's engines uncovered subsurface ice as it landed (NASA/JPL/University of Arizona, Max Planck Institute)

hydrochloric acid, and other sulfate and chloride species, interacting with the soil and the atmospheric dust."

Modern meteorology on Earth benefits from centuries of observation. The comprehension of Martian weather and climate will require long-term data. That's something that is just beginning to happen, contends Steve Lee. "We had some ideas that if you look in a particular area on a particular day of the year, that you can be reasonably certain that something is going to happen in the atmosphere – whether it is a dust storm or clouds that will form or whatever, because the information so far seems to be that it's very repeatable from year to year. But not always – and that's where our understanding breaks down, because we don't have a long enough database to really start making those conclusions."

"Now, certainly doing all these observations in a repeatable fashion where you get – month after month – some of the observations and then just do it continuously over multiple years is a big help. Tying back into the Mars Global Surveyor, which gave us 10 Earth years of observations, so roughly 5 Mars years, we're approaching a Mars decade, which is starting to become an impressive database."

"Moving back in time, we've got the Hubble observations we did in the 1990s, which sort of give you the tie in between Earth-based telescopes and the orbital observations. We go from that back to the Viking missions in the 1970s and Mariner 9 in the early 1970s, so we're starting to get some observations that give us approaching 40 years of observations with spacecraft to do these things. That's particularly been useful for things like polar observations, where we see things changing in the south polar caps in particular with MRO, and there are a minimal number of observations from Mariner 9 that were high enough resolution, maybe a little bit better than 100 m per pixel, where you can at least see the gross morphology of these features and say, well, yeah, it was doing the same thing back then, or it was doing something different. So the puzzle pieces are starting to fit together much better."

One area under intense study is the connection between the Martian surface and atmosphere. The idea that a planetary surface is directly tied to its weather is widely accepted today, but it has not always been so, says planetary geochemist Ben Clark. "Some atmospheric scientists didn't recognize that for quite a long time. Earth has a small amount of carbon dioxide in its atmosphere, but if you could take all the CO_2 back out of the limestone on Earth, you would have an atmosphere that's about as thick as Venus's atmosphere. One thought about why Venus has this greenhouse problem is that its carbon dioxide never got taken up and sequestered in rock."

Phoenix's Peter Smith sees a strong relationship between Martian soil and sky. "We think that the soil produces a sort of local weather in terms of the humidity. We found perchlorate [a highly oxidizing agent] in the soil and other minerals that are very much attracted to water, so you end up with a cycle where the atmospheric water drops out at night, at least in the late summer, and saturates the soil. In the daytime when it heats up, the soil releases the water back into the atmosphere, and you get these daily cycles. That includes even snow."

MRO researcher Steve Lee adds, "There's sort of respiration (though that's anthropomorphizing it). There's an exchange of water vapor in and out of the regolith, over the course of a day, or when it's getting colder. It probably goes into the top layer of the regolith, and when it warms up a little bit it sort of gets expelled slightly. But you can get an enhancement of humidity near the surface."

The Mars rovers Spirit and Opportunity were born under the umbrella of NASA's "Follow the Water" plan. But Ben Clark sees a shift in research. "The strategy of following the water was, I think, a good one. It tells us about different climates Mars has had. Did it have rainfall? Is life a possibility, past or present? But the next step has been to follow the carbon." Carbonates provide constraints on the extent of an ancient Martian atmosphere, and organic carbon is relevant to the ultimate questions about life.

Still, much remains to learn about Martian weather, past as well as present. We still do not fully understand the relationship between the polar caps and an atmosphere that ebbs and flows with the seasons. The dynamics of global dust storms and layered records in polar ices have much to tell us about global warming and climate change. Even the wind patterns and

Martian sunsets, at once familiar and alien, many make future travelers homesick (Spirit rover NASA/JPL)

pressures continue to mystify us. For scientists such as Clark and Smith, it is merely sauce for the goose.

From the standpoint of human exploration, Mars is the most accessible planet in our Solar System. It is second only to Venus in distance, and its thin atmosphere makes a safe return to Earth much more feasible than a return launch through the dense atmosphere of Venus. The weather of Mars is by far the most like home, with snowfall, reasonable temperatures, and similar seasonal and day/night cycles. Future visitors will, in some ways, feel right at home looking across the desert landscape at a sky filled with not-too-alien winds.

A Question of Life

The Martian past haunts exobiologists in their search for life on other worlds. With all the similarities between Mars and Earth, they ponder the possibilities of life on Mars, past or present. Life on Earth requires water in liquid form. The Phoenix lander found water-ice just under the surface at its northern landing site. With Mars's past axial wanderings, Peter Smith wonders what Phoenix's site would be like under those past conditions. "This (axial wobble) is a periodic effect, so you expect this to occur many times in the future as well as the past. Does that create a habitable zone at those times? Perhaps it does. You can build a good case for it. Could microbes exist that are active for a few thousands of years and then have to go dormant for a long period of time? I don't know. It's a pretty tough life up there. So I don't know that the place we were at would be my best choice for where you'd find life. It was a good choice to look for polar conditions, but if you're looking for life, there are probably better places. Cracks that are communicating with subsurface heat sources, or volcanic activity that might be near-surface, those have got to be better environments. Mars is a big place." Smith feels that conditions on Mars would not have to change significantly to create habitable zones. "At this time you've got water and vapor closely interacting with soils. Certainly in the past when it was warmer and there was more water available, conditions were getting very close to where you might have thin films of liquid that were very stable in the soil. It doesn't take much in terms of a thin film to have a biologically active environment. You can even have communities of microbes associated with films of water as thin as the wavelength of light, 500 nm. If you're releasing the polar caps into the atmosphere, the pressure's going up, clouds are getting denser, it can change the climate dramatically."

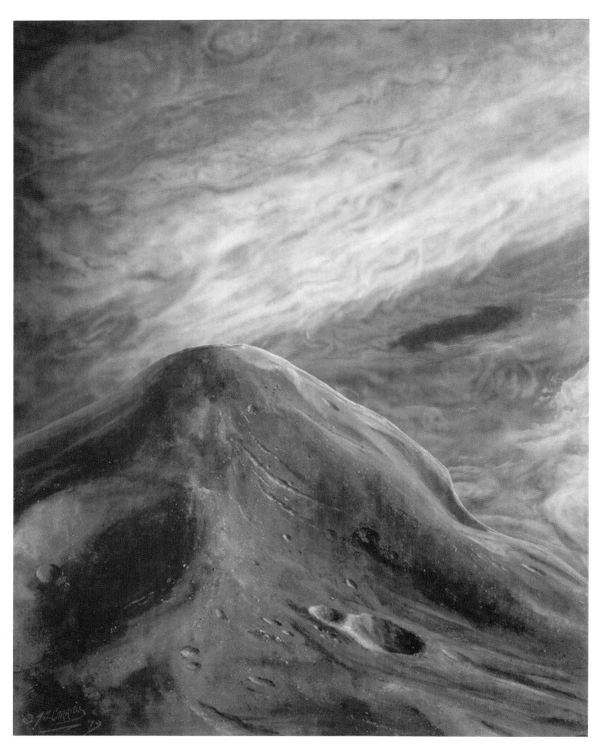

A visitor to Jupiter's tiny moon Adrastea would be treated to a spectacular parade of ever-changing clouds on the king of worlds. Adrastea orbits just 57,200 km above Jupiter's cloud tops; the gas giant would span as far across the sky as 280 full Moons in Earth's sky. Note the dark "hot spot" opening in the cloud deck at right center (© Michael Carroll)

Chapter 7
Jupiter

A mong the outer planets, we find the coldest temperatures and fiercest winds. The gas giants – Jupiter, Saturn, Uranus, and Neptune – offer us the most alien skies in our Solar System. Here, where the Sun glows like a distant ember, storms large enough to swallow the entire Earth rage for decades. It is a region of extremes, of bitter darkness and numbing cold. Earthly gases become liquid or ice. Snows of ammonia- and methane-ice crystals drift from lightning-laced clouds. In other places, rains of methane and ethane fall across frozen landscapes.

THE BIG PICTURE: GAS GIANT ANATOMY

The gas giant planets are worlds of weather. Unlike the terrestrial planets, their surfaces are not ice and rock, but rather gases and clouds, cryogenic rainstorms, and streams of supersonic winds. Their cores are far more dense than Earth's core, but as one moves out from the center of each planet, the environment transitions from a rocky and metallic solid to a liquid, then to a gas. There is no abrupt delineation between solid and gas, no place to stand on these cosmic behemoths. Because there is no surface from which to measure altitude, scientists instead use pressure as a reference point, with 1 bar (equivalent to Earth's atmosphere at sea level) as the starting point.

With no solid surface to prop up the air, the atmospheres of the gas giants are self-supporting. The great pressures of deep atmospheres act as a foundation against the force of gravity, keeping the atmospheres in balance with the pull of the planet. This balance is called hydrostatic equilibrium.

Perhaps more than anything else, it is the scale of the gas giants that boggles the mind. In essence, our Solar System consists of the Sun, Jupiter, and a few incidental bits. Jupiter itself is more massive than all the other planets and their moons put together. Over 1,300 Earths would fit inside of Jupiter. Though far less dense, Saturn is not much smaller, and its ring system would stretch two-thirds the distance from Earth to its Moon. Uranus and Neptune are twins in size, both encompassing as much volume as 60 Earths. Although the meteorology of Uranus is subtle, Neptune's troposphere births cyclones the size of respectable terrestrial planets.

Jupiter, Saturn, Uranus, and Neptune have reducing atmospheres, that is, atmospheres dominated by hydrogen and helium, the ancient building blocks of the Solar System. Their gravity is so strong that they were able to hold on to even the lightest primordial gases that initially gathered to form them. The ratios of helium and hydrogen within both Jupiter and Saturn are similar to the makeup of the Sun itself. Uranus and Neptune have evolved slightly different mixes of gases.

Taken as a planetary quartet, the giants display common patterns and fundamental differences. Their atmospheres, the places where weather occurs, are mere films less than half a percent of each planet's diameter. But these veneers generate dynamic, powerful, and varied weather. All four

M. Carroll, *Drifting on Alien Winds: Exploring the Skies and Weather of Other Worlds*,
DOI 10.1007/978-1-4419-6917-0_7, © Springer Science+Business Media, LLC 2011

Earth compared to Jupiter's Great Red Spot (© Michael Carroll)

Voyager 2 imaged Jupiter's Great Red Spot in 1979 (left). Nearly three decades later, the Cassini spacecraft sent another view (right). Note subtle changes in the bands, zones, and Great Red Spot (NASA/JPL)

worlds have belts of clouds encircling them parallel to the equator. The cloud bands are wracked by incredibly strong winds, but they remain stable within their latitudinal paths. On Earth, bands of clouds are broken up by great storms at temperate latitudes from the tropics to the arctic circles. Continents block airflow, causing pressure waves that also mix things up. Storms come and go on the order of days or, in the case of cyclones and monsoons, weeks.

But the bands and giant storms of the gas giants may last for decades or even centuries. Astronomers have observed Jupiter's Great Red Spot (GRS) for nearly four centuries.

The cloud banding on the gas giants is most obvious on Jupiter and Saturn, where the Sun shines most directly on the equator. Despite its distance from the Sun, Neptune's bands are nearly as well defined. Even Uranus, whose axis is tilted so that the planet essentially rolls around the Sun on its side, has subtle banding. To California Institute of Technology's planetary atmosphere researcher Andy Ingersoll, this is significant. "At Uranus during the Voyager encounter, the Sun was almost directly over the south pole. Despite this very strange orientation of the Sun, the cloud patterns made it look just like a tipped-over version of all the other giant planets. The jet streams and the cloud bands follow circles of constant latitude. It's as if you took Jupiter, tipped it over, and painted it out so it didn't have as many details. Basically this says that the Sun does not control the orientation of the cloud patterns and jet streams. It's the rotation axis that controls it."

The banding of gas giant clouds gives these planets their distinctive appearance. Eastward and westward winds segregate the clouds into patterns of dark belts and light zones. Earth has something like these in each hemisphere. The trade winds form a westward air current at latitudes near the equator, while the jet stream wanders in eastward currents at mid-latitudes. Neptune's belts are arranged similarly, with two major currents in each hemisphere. Voyager saw only one hemisphere of Uranus, but ground-based data reinforces the suggestion that similar bands play across its globe. Jupiter has five or six belts and the same number of zones in each hemisphere. Zones tend to be sinking air masses, while the belts contain ascending air. Saturn's belts are harder to see, as they float beneath golden hazes, but it is clear that the ringed giant has a similar number of belts with much higher winds than those found on Jupiter.

All four gas giants spin rapidly. Jupiter's day lasts a scant 9 h 55 min. Saturn takes about 10½ h to rotate, but the planet is less dense than Jupiter, so its spin flattens its poles substantially. Uranus and Neptune turn once each 17¼ and 16 h, respectively.

The air of the gas giants is a rich brew of complex chemistry. Gases in the atmosphere transform into organic material and heavier gases. This transformation takes place at the hand of forces that include lightning, rising currents from interior heat, cloud formation (condensation),

On December 29, 2000, the Saturn-bound Cassini/Huygens spacecraft took this elegant portrait of Jupiter. Note how opposite-trending winds shear the clouds at the edges of belts (dark areas) and zones (light bands) (NASA/JPL)

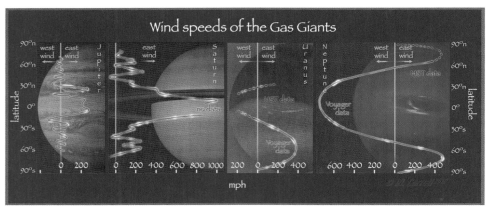

Winds of the gas giants vary dramatically from one planet to another. Neptune has the greatest difference in speeds from westerly to easterly. Voyager only saw one hemisphere of Uranus, so data in the north comes from Hubble Space Telescope observations. The same is true for the northernmost region of Neptune. Note that the zero mph vertical line indicates the actual spin rate of the planet's core, identical to the speed of the atmosphere's rotation. Saturn's line has been moved to the east by 40 m/s compared to Voyager data to compensate for new information on the planet's spin rate (see Chap. 8) (© Michael Carroll)

Interiors of the gas giants Jupiter and Saturn are so massive that their cores are surrounded by liquid metallic hydrogen covering a soup of super-pressurized water, ammonia, and methane. Pressures at the center of the smaller Uranus and Neptune do not reach those levels. The top 1% of each planetary slice – thinner than the peel of an apple – is where all the weather occurs (© Michael Carroll)

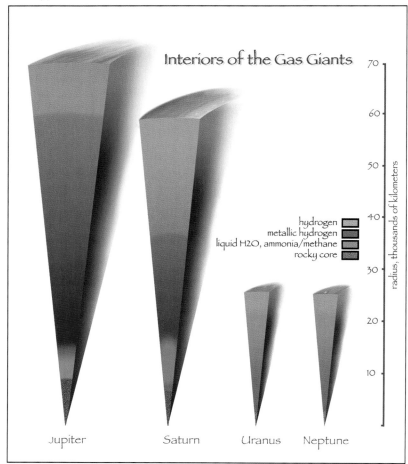

and photodissociation, the process in which sunlight splits molecules of gas. As a result of this titanic chemistry experiment, the clouds of Jupiter and Saturn are painted in rich oranges, tans, browns, and blues. Farther out, methane tints Uranus and Neptune toward the blue end of the spectrum. Neptune's clear air reveals a rich teal cloud deck, while hydrocarbon hazes tint Uranus to a pale shade of blue-green.

While the outer two blue worlds have at least 15% helium in the air (similar to the Sun), something has pulled the helium from the upper atmospheres of Jupiter and Saturn. Scientists believe that an ocean of metallic hydrogen surrounds the massive cores of these two planets. Helium drops may form within these deep seas, slowly raining down toward the center of the planets. On Uranus and Neptune, the core pressures are not high enough to force hydrogen into a liquid metal form, so this helium rainfall does not happen there.

Models indicate that there are three layers of clouds on each of the gas giants. On Jupiter and Saturn, the highest deck consists of ammonia. Below this floats a mixture of ice crystals, part ammonia and part hydrogen sulfide. Underneath these cloud decks, mists of water-ice or water vapor simmer in the depths. Neptune and Uranus also appear to have their cloud decks arranged similarly, but their colder temperatures do not form high ammonia clouds. Instead, brilliant white clouds of methane float above a blanket of lower clouds. The lower deck may also be methane. Deeper water clouds may contain ammonia, but if so, they are hidden beneath the middle deck.

THE STUFF OF JUPITER

In 2001, the alternative rock group Train brought its song "Drops of Jupiter (Tell Me)"[1] to the top 40 for 38 weeks. Lead vocalist Pat Monahan sings, "Now she's back in the atmosphere with drops of Jupiter in her hair…" Forgetting, for a moment, that anyone who had gotten drops of Jupiter in their hair would have a lot more to worry about than their coiffure, what would those drops consist of? And where would they come from?

After decades of probes and centuries of observation, we are left with a fairly good idea of what we would see if we drifted down through Jupiter's sky. In the lowest part of the stratosphere, the sky around us is a deep blue to purple. Clear air, laced with poisons of hydrocarbon hazes, refracts sunlight

Voyager view of the Great Red Spot (NASA/JPL)

1. "Drops of Jupiter (Tell Me)," words and music by Pat Monahan, James W. Stafford, Scott Michael Underwood, and Charlie Colin, © 2001 EMI April Music Inc., Blue Lamp Music, EMI Blackwood Music Inc. and Wunderwood Music. All rights for Blue Lamp Music controlled and administered by EMI April Music Inc. All rights for Wunderwood Music controlled and administered by EMI Blackwood Music Inc. All rights reserved. International Copyright Secured, used with permission. Reprinted by permission of Hal Leonard Corporation.

toward the blue due to Rayleigh scattering, just as air does in Earth's skies. Below us waft delicate white clouds of ammonia ice crystals riding on updrafts. As we descend into the troposphere, we reach this patchy cloud deck. Pressures increase and temperatures become warmer. At the top of the bright ammonia clouds, the pressure is about a tenth that of Earth at sea level – 0.1 bar – and the temperature is −243°F, on the rise. Ammonia snows fall from the clouds, headed for deeper regions.

Coming up beneath us swirls the rich rust-brown cloud deck that gives Jupiter its dark banding. This layer of ammonium hydrosulfide cloud, with a base about 50 miles below the ammonia deck, may be laced with rich organic compounds generated from energy farther below. This energy includes heat and radiation flowing from within Jupiter, as well as complex chains of amino acids cooked from the clouds by powerful lightning. As the skies darken, storm clouds shimmer in the glow of distant lightning. Compared to Earthly skies, lightning is fairly rare here, but a Jovian bolt packs enough force to power a small town for days.

Hurricane-force winds buffet clouds around us into eddies and whirl-pools of air. Oval storms the size of Earth's continents spin like hungry maws, eating away at the clouds around them. Banners of feathery mists stream for hundreds of miles along powerful jet streams. Spots of rich color circle each other, intermingle, and merge and dissipate in a ceaseless interplay of color, motion, and changing texture. The ammonium hydrosulfide blanket of cloud spreads out in a jagged plain to a horizon many tens of times the distance of Earth's own. Jupiter is so large that there is no sense of objects disappearing over the horizon. Massive towers of clouds simply fade away with distance. Here and there, pale blue-gray clouds break through the ruddy plain, water clouds boiling up from below. The water cloud deck is the lowest, floating nearly 100 miles below the highest ammonia cirrus. Here, temperatures rise above the melting point of water, allowing water vapor to billow into clouds. Underneath this cloud layer, rain falls into an eternal night of crushing pressures and searing temperatures, as the hydrogen making up most of the air around us transforms into a liquid, and then into a fluid metal.

Jupiter's vast, brightly colored bands of cloud have long mystified observers. Early astronomers guessed that Jupiter's Great Red Spot, which undulated and varied in size and shape, was the plume from a titanic volcano. The truth was just as strange, and the GRS still stands today as the largest cyclone in the Solar System. Scientists have been trying to figure out those Jovian clouds for centuries. Andrew Ingersoll has been studying planetary atmospheres for over three decades. The rich color of Jupiter's clouds has yet to tell us their true composition, he says. "Color on all the giant planets is one of our embarrassing failures. Take Jupiter's color, for example: sulfur compounds are sort of reddish, and organic compounds can be reddish. Jupiter is a pale pink-orange thing, and we don't really know whether we're looking at sulfur compounds or organic compounds or even phosphorous red." Unlike gases, which have a unique spectroscopic fingerprint,

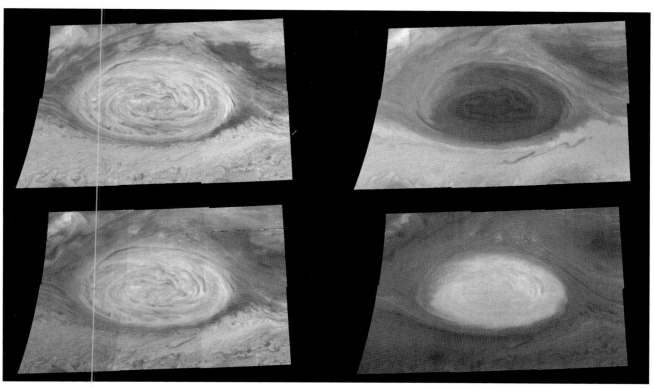

Comparison of Jupiter's Great Red Spot at four wavelengths. These mosaics (six frames each) show the appearance of the Great Red Spot in violet light (415 nm, upper left), infrared light (757 nm, upper right), and infrared light within both a weak (732 nm, lower left) and a strong (886 nm, lower right) methane absorption band. The images were taken within minutes of each other. Reflected sunlight at each of these wavelengths penetrates to different depths and is scattered or absorbed by different atmospheric constituents before detection by Galileo

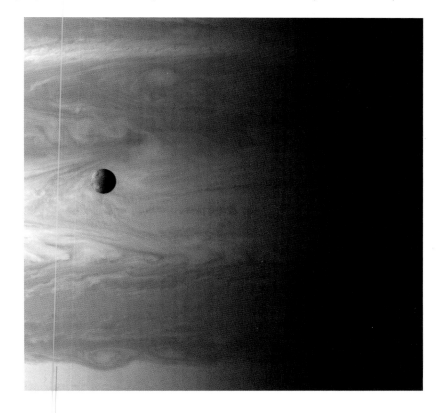

The Saturn bound Cassini spacecraft snapped this portrait of the moon Io floating in front of Jupiter (NASA/JPL)

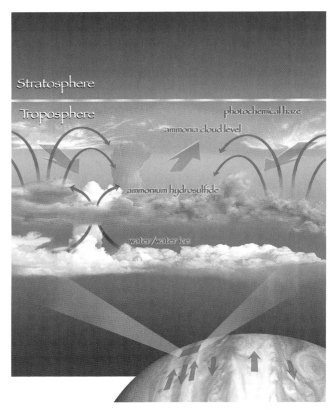

A slice of the planet shows the layered arrangement of Jupiter's three distinct cloud decks. The deep browns and reds of the planet come from a layer of ammonium hydrosulfide clouds, while bright ammonia ice clouds condense as hot air upwells from below. Farther down, a blanket of water clouds occasionally punches through the middle deck. As on Earth, clouds and weather tend to remain in the troposphere (© Michael Carroll)

making them easy to identify, solids (like clouds, which are little particles of solids), don't have a clear fingerprint that can be measured by remote sensing. Researchers believe the bright highest-altitude clouds on Jupiter consist of ammonia. A mid-altitude cloud deck is made up of ammonium hydrosulfide crystals, a combination of ammonia and hydrogen sulfide. Ingersoll's mystery compounds tint this deck. Still deeper, water clouds may lurk in the warmth of high pressure at about 5 bars.

Jupiter's atmosphere, like all other atmospheres, strives for balance, not only in temperature but also in chemistry. Because of the way molecules combine, and because hydrogen is the most common component of the atmospheres of the gas giants, carbon should be rare, having combined with hydrogen to make methane. In the same way, oxygen should exist only as part of water and nitrogen as ammonia. But Jupiter's atmosphere is dynamic; these elements have been seen, and so have others that do not remain stable for long in the presence of hydrogen. These include ethane, acetylene, phosphine, and carbon monoxide. It may be that the turbulent atmosphere brings gases up from deep in the planet before they can combine with others. If all Jovian clouds were in chemical balance, they would be white. The colorful face of Jupiter owes its visage to chemical reactions as molecules try to find bal-ance. After all, that is what we are all after.

The colorful storms on Jupiter rage for long periods at mind-boggling scales. Their drivers are a complex mix of wind shears, solar energy, planetary spin, and updrafts. Those elements came into focus with the Voyager and, later, Galileo encounters. "The fundamental surprise was how turbulent the atmosphere was given that the large structures are so permanent and everlasting. Another surprise was that the small scale was just a mess, with things changing every day. Later, we were surprised to find that the eddies were not just parasitic, but were actually pumping up the large motions."

The concept that small-scale weather could feed and transform large meteorology (like global jet streams) was one familiar to meteorologists. In the forties and early fifties, scientists began to embark on a careful statistical analysis of Earth's weather. They realized that Earth's jet streams were actually driven by small-scale turbulence. It seemed paradoxical, says Ingersoll. "It wasn't turbulence in a random sense. It was organized turbulence.

It's the big frontal cyclonic systems that are the weather at mid-latitudes. They realized that those big cyclones and anticyclones were actually maintaining the jet stream, and *not* the reverse. It wasn't that the jet stream was there and it spun off these eddies. It was as if the eddies were maintaining the jet stream." Researchers started looking for similar phenomena in the motions of the clouds on Jupiter and found the same situation. "There's only one eastward jet stream in each hemisphere on Earth; on Jupiter there are six or seven. We found that each one of those was being pumped by the eddies. Jupiter is a better example of this pumping than even Earth."

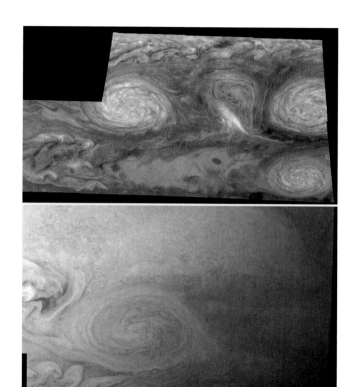

Another key element to Jovian weather concerns heat flowing from Jupiter's interior. This escaping energy sets up a situation quite different from terrestrial weather, Ingersoll says. "Just as the longevity of the weather was totally different from Earth, also the energy balance from equator to pole was totally different. On Earth, as we all know, it's really much colder at the poles because the Sun heats the equator, and the infrared radiation leaves an excess at the equator that has to be transferred down toward the poles. That's what the weather does. On Jupiter and then on Saturn, the temperature at the equator is just about the same temperature as at the poles. We even found this to be true on Uranus and Neptune. Up in the stratosphere, it is warmer in the summer hemisphere at the equator, more intuitive based on Earth intuition, but the bulk of the atmosphere – the troposphere – is just about at the same temperature. That difference, we're still trying to sort out." Two major theories have been put forth. One is that the internal heat – which giant planets have – is forced to come out preferentially at the poles and make up for that balance. The atmosphere itself does not transport heat efficiently from the warm equator to the cold poles. Ingersoll likens the polar heat escape to a thermostat; it keeps the temperature constant.

Jupiter's small-scale turbulence took early researchers by surprise. Top: Image of Jupiter's tumultuous white oval storms, taken by Galileo (JPL). Bottom: This image of Jupiter's "Little Red Spot" – a smaller version of the GRS – was snapped by NASA's New Horizons spacecraft on its way to Pluto (NASA/Johns Hopkins University Applied Physics Laboratory/Southwest Research Institute)

A second leading theory proposes that atmospheric heat transfer is much more efficient on the giant planets than it is on Earth, so that it is able to reduce the temperature differences to very small, unmeasurable values. But if the internal heat flow is one engine and the horizontal heat transfer in the atmosphere is the other engine, which one has the more sensitive throttle? Ingersoll advanced the argument that it is the internal heat that has the more sensitive balance. "If you clear the poles off just by a fraction of a degree, you get a big transfer of energy from below into the poles. If you cool a fluid at the top, convection will bring heat up from below. Sideways transfer of heat

Jupiter seen in infrared. The planet puts out more energy than it receives from the Sun. Note the bright "hot spots" north of the equatorial zone (NASA/JPL/IRTF)

The interior of the gas giants may be arranged as stacked counter-rotating disks (© Michael Carroll)

requires more energy. On Earth we have 30 or 40°C difference between equator temperatures and the pole. On Jupiter it's maybe a fraction of a degree difference between equator and pole."

The internal heat of the gas giants is actually fairly tame, says planetary scientist Anthony Delgenio of NASA's Goddard Space Flight Center. "You talk about a place like Jupiter, and you talk about this strong internal heat source, but that strong heat source on Jupiter is a few watts per square meter of area. Compare that to how much heat is leaving Earth's surface, and on Earth's surface it's over 100 W/m². This supposedly strong internal heat source on Jupiter is a very relative thing. It's strong compared to the sunlight that's entering, but in absolute terms it's very weak."

Despite its comparatively low output, Jupiter's internal heat results in dramatic weather. Scientists are trying to understand the alien meteorology by the use of computer modeling. Jupiter is so complex that many models must simplify the variables seen in the real world, says Andy Ingersoll. "There are models that really treat Jupiter's atmosphere as if it was a big Earth. Instead of this big atmospheric layer resting on continents or oceans, it's just resting on more atmosphere underneath it, but the idea is that the deep atmosphere [of Jupiter] is quiet and not doing anything. The other school of thought says that the internal heat is generating motions all the way down, so that what we see at cloud top is just a manifestation of three-dimensional motions that extend into the fluid interior of the planet. There are computer models of both phenomena and both satisfy the available data, but we don't have the 3D data to really choose. The Galileo probe went in at one place and the wind did increase with depth. That would favor a deep flow model as opposed to the thin weather layer model, but the probe only got down to 20 bars, which is just a fraction of a percent of the way down to the center of the planet, so the issue isn't solved yet."

Jupiter's belts and zones may be more than skin deep. Atmosphere acts as fluid. Fluids in a rotating sphere line up with the axis of rotation. All the gas giants may be arranged as a series of stacked disks, each rotating at its own speed. Zones may simply be the surface effect of these rotating cylinders.

In an effort to understand the layered structure of the giant world, NASA launched the Galileo mission in October of 1989 aboard the space shuttle Atlantis (see Chap. 3). Its circuitous route brought it to the Jovian environs in December of 1995 after two flybys of Earth, one of Venus, and two asteroid encounters. Just 150 days before orbital insertion, the massive orbiter released its atmospheric probe. The probe's goals were ambitious:

1. To determine the composition of Jupiter's atmosphere
2. To chart the structure of the atmosphere down to the 10 bar level, and map the vertical locations of the cloud decks
3. To investigate the composition of particles in the clouds
4. To search for lightning
5. To monitor Jupiter's heat flow
6. To measure the charged particles flowing into Jupiter's upper atmosphere from its magnetosphere

Galileo found that Jupiter contains helium in nearly identical proportions to that of the Sun. On its descent, it encountered 400 mph winds that buffeted the coffee-table-sized probe. Experiments found lightning to be more rare than on Earth, with 10% of the discharges of a comparable terrestrial area. Most baffling was the lack of water, far less than scientists had predicted, says Andy Ingersoll. "It never really got to the water, although it did get into a region where water was increasing with depth. Water is another big unknown, a missing link in all of our studies. We have not successfully measured the water abundance of the deep atmosphere or, therefore, of the planet itself. It's deep down because water is frozen out at the upper levels where it's very cold."

Galileo sailed through much clearer air than expected. Both the lack of water and lack of clouds mystified the science teams on Earth until the Hubble Space Telescope imaged Galileo's entry site on Jupiter. The orbiting observatory revealed that Galileo had fallen into what scientists refer to as a hot spot, an opening in the clouds. Says JPL's Kevin Baines, "It's like Murphy's Law is everywhere: We had cloud-measuring instruments; we wanted clouds, but we happened to go into this clear place. It was all dictated by celestial mechanics. The probe had to go in at a certain latitude, and whatever was there longitudinally was what you got. It just happened to fall into the brightest hot spot on the planet at the time. So we saw very few water clouds." For a few days, researchers thought they would have to revamp their models of the Solar System because the probe found no water. Water content is tied to how much oxygen is in the air, so determining water vapor content was one of the primary goals of the probe's mission. Oxygen is the most abundant element in the universe next to hydrogen and helium. Without information on the planet's oxygen levels, planetologists are missing an important piece of Jupiter's composition. To Andy Ingersoll, "the water is still the great mystery for me. How much water is contained within Jupiter? Not only does it complete our inventory of oxygen and hydrogen in the Solar System, but it's important to meteorology. We've seen magnificent lightning [at Jupiter]. Thunderstorms appear to feed all other meteorology, including the oval storms; they're at the bottom of the food chain. These storms are governed by water."

When the Galileo investigators realized where the probe had fallen, they scrambled to make a "Plan b." While the probe didn't get the information on the water content or, more specifically, the oxygen content, of the planet, the mother ship got data that brings a close approximation. The Galileo orbiter was able to prove that there was more water elsewhere. Baines observes that, "If that probe had just fallen 3,000 km away (and 3,000 km is a very small distance on Jupiter) it would have found lots of water."

Galileo's hot spot landing site is interesting in its own right, says Andy Ingersoll. "It's definitely an anomalous region. That particular latitude has wave-like structures. At that latitude these hot spots are arrayed around the planet." The string of clearings stretch for thousands of miles each, and they ring the equator in the north. Scientists wonder why the hot spots congregate only at 6° north latitude. Why would they not also exist at 6° south latitude? Baines says this is not the only asymmetry found on Jupiter. "There's the same kind of asymmetry with the red spots: why are they only in the south? There seem to be eight to twelve of these northern hot spots

A hole in the cloud deck of Jupiter, just left of center. This hot spot is similar to the area into which the Galileo probe descended. (JPL)

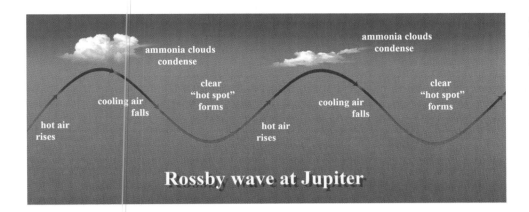

ammonia clouds
condense

ammonia clouds
condense

clear
"hot spot"
forms

clear
"hot spot"
forms

cooling air
falls

cooling air
falls

hot air
rises

hot air
rises

Rossby wave at Jupiter

A Rossby wave might explain the ring of hot spots encircling Jupiter north of its equator (© Michael Carroll)

around the planet at any given time." Adjacent to the hot spots, and to the south, ammonia clouds form in tufts all the way around the planet. On Jupiter, ammonia clouds tend to form where there is uplifting. Baines believes that researchers may be seeing a 3D wave called a Rossby wave. "A Rossby wave may link them all together, so that what we're seeing with the hot spots and ammonia clouds is a manifestation of the same phenomenon. The wave goes up and all of the ammonia comes out of the air, and the wave descends. It's dry now, so it creates a hot spot. As it comes back up again, it forms another cloud."

This atmospheric roller coaster might also explain the lack of other constituents in the probe's results, says Andy Ingersoll. "If a hot spot is a trough of the wave, then it pushes all the volatiles down to deep levels so the upper levels are dried out, and that's basically what the probe saw. It finally saw the ammonia, but at much deeper levels than we expected, and it finally saw the hydrogen sulfide, also at much deeper levels."

Although pressure and temperature dictates that the bright upper clouds are made of ammonia, only newly formed clouds seem to have the spectral signature of the gas. Theorists propose that as ammonia boils up from below, it remains pure for 3 days to a week. Then, ammonia molecules become blanketed by hydrocarbons from the organic hazes in Jupiter's upper atmosphere.

Understanding the weather phenomena on Jupiter is critical to the understanding of Earth's meteorology, says Andy Ingersoll. "If all your theories are based on Earth, you can't really test them, so you go to another planet where you can play 'what if' games and learn things." On our own world, storms are present everywhere, and jet streams dominate the atmospheric circulation. Jupiter is an exaggerated caricature of terrestrial meteorology, a natural laboratory where atmospheric scientists can study the nature and interplay of the intense jets and severe atmospheric phenomena.

Anatomy of a Big Red Blob

The Great Red Spot has raged across the face of Jupiter for at least three centuries. During that time, it has changed color, shape, and latitude. At the time of the Pioneer encounters in 1973 and 1974, the cyclone was surrounded by a dense white band. The spot itself was a fairly uniform color. In the time between those encounters and the arrival of Voyager 1 in 1979, the white band had broken up into undulating waves of cloud, and the spot itself had a more inconsistent and varied pattern of inner swirls. In the brief months leading up to Voyager 2, the spot became more uniform and rich in color, resembling its appearance during the Pioneer flybys. By the time Galileo settled into orbit in 1995, the GRS had again faded somewhat, and clouds around it took on new forms.

The Great Red Spot rotates anticyclonically, that is, counterclockwise. The storm is actually a vast dome of cloud. Its form has the appearance of a hurricane, but unlike hurricanes, the center of the GRS is a high-pressure region. Clouds flow out from the center in an S-shaped pattern. The cyclone acts as an obstacle in Jupiter's alternating winds, forcing streams of cloud to flow around it. Observers clocked winds within the spot at up to 600 kph.

Ammonia clouds are associated with dramatic updrafts in active areas. One active site is a turbulent region to the northwest of the Great Red Spot. This area is known as the Turbulent Wake Region (TWR). There, the Great Red Spot acts like a rock in the middle of a stream. Jupiter's jets trend in the east-west direction. The Great Red Spot blocks the streams, forcing them around it. As the streams get diverted north, momentum tries to carry them south again, but they crash into the streams of air on the other side. Violent thunderstorms form continually in the Turbulent Wake Region.

Waves in the Air

One force that constrains and sculpts atmosphere is called a Rossby wave. A Rossby wave uses the Coriolis effect as its "restoring force." Any wave has a restoring force that tries to pull it back to center. A stretched rubber band, when let go, oscillates back and forth because of a restoring force that tries to pull the object back to center, but it overshoots the center and goes the other direction. As the rubber band stretches, it creates a force in the other direction, and it overshoots again. The Coriolis force dampens that effect.

Saturn nights are spectacular. The rings arc across the evening sky beyond towers of ammonia and water clouds. At the horizon, the rings appear to bend in the dense air (© Michael Carroll)

Chapter 8

Saturn and Titan

The Lord of the Rings, Saturn, receives one fourth of the solar energy that Jupiter does (one hundredth that of Earth). Predictions held that its meteorology would be less vibrant than Jupiter's because of its distance from the Sun. Telescopic views show muted belts and zones, tame versions of those on Jupiter. But in light of the Voyager flybys, Saturn's weather turned out to be just as tumultuous, and in its own unique style. With revelations brought to us by Voyager, JPL's Kevin Baines was mystified, then amused. "Saturn is very subdued and lovely, but on the outside only. Saturn is just as dynamic and exciting as Jupiter once you get below the haze. The gravity is low and the particles of haze can hang around for a long time." The longevity of hazes and cloud features is counterintuitive to those who study Earth's weather. Caltech's Andy Ingersoll counsels that we cannot always draw parallels between terrestrial meteorology and that of the outer worlds. "There's a theme here: Earth analogy didn't always work for the giant planets, and that's always part of the fun, part of the intellectual progress."

Ingersoll describes the fog that is responsible for Saturn's glorious golden tint. "The haze is probably formed when ultraviolet light from the sun knocks a hydrogen off of a methane molecule. The methane goes around, and instead of grabbing another hydrogen it grabs another fragment of methane molecule, and they form a higher hydrocarbon with more than one carbon atom like acetylene or ethane. You can build up longer-chain hydrocarbons and those start to condense. Pretty soon you have haze. So essentially, we have smog."

Voyager's robot eyes showed waiting scientists a planet torn by supersonic winds, yellow-white clouds sheared across dark bands like the banners of racing chariots, a filigree of mists and fogs curling and twisting across white whirlpools and canyons of cloud, all tinted by golden hydrocarbon hazes. A band of powerful thunderstorms encircled the southern hemisphere, gaining the moniker of "Thunderstorm Alley." Tantalizing geometric forms ringed the poles.

The planet's winds were among the fiercest in the Solar System. The variety of Saturn's weather was unexpected. Nearly three decades later, the Cassini Saturn orbiter simply reinforced the alienness of Saturn compared to its giant sibling.

To many, Saturn has turned out to be the most photogenic planet in our Solar System, says Cassini Imaging Team Leader Carolyn Porco. The vast majority of images have been targeted for scientific, not aesthetic, reasons, but "they just happen to also be these beautiful graphic designs. No one ever stopped

Saturn is subdued and lovely…on the outside (NASA/JPL/Space Science Institute)

M. Carroll, *Drifting on Alien Winds: Exploring the Skies and Weather of Other Worlds*, DOI 10.1007/978-1-4419-6917-0_8, © Springer Science+Business Media, LLC 2011

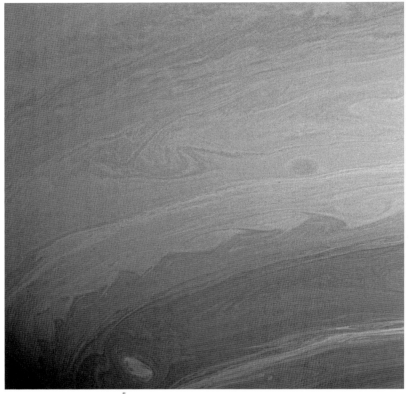

Undulating clouds paint Saturn's face in subtle swirls in this enhanced Cassini image (NASA/JPL/Space Science Institute)

The striped banding of Saturn's belts and zones shows through shadows cast by the rings, seen at left with the Sun behind them (NASA/JPL/Space Science Institute)

to think that when we take a picture of the rings through the atmosphere of Saturn you're going to see the rings bend. Things like that are amazing. The big popular mosaic taken in eclipse highlights rings you can't otherwise see, so the ring scientists wanted to do that, but they were going to take half of it. I kept saying, 'no, no, you want to take the full mosaic for PIO [the Public Information Office].' Porco is glad the team took the full panorama. It was among images cited by the committee that awarded her the prestigious Lennart Nilsson Award. "All the structure in the rings took everybody by surprise; the spokes, the ringlets, all of it was just amazing. Just imagine: you've got a sheet of debris 30–60 ft thick. Imagine you're in a shuttlecraft, and it's like you're going to the Moon; the rings would fit nicely between Earth and the Moon. You're skimming over this sheet of debris as far as your eyes can see. If you're just above it, it will look like it's going out to infinity. You come upon this wall of debris that's 2 miles high, just rubble that's been thrown up." The gravitational interplay between rings and moons gives rise to these walls of ring particles, along with undulating waves, gaps, and clumps.

Cassini also added to our knowledge about Saturn's retinue of mid-sized moons, providing stunning views of two-toned Iapetus and geyser-laden Enceladus. But Saturn's atmosphere continues to challenge and awe researchers. Saturn's powerful winds stand as one of the most difficult challenges to observers. The reason – planetary scientists simply do not know with certainty the length of Saturn's day.

Since gas giants have no solid surface, the weather provides the only visible features with which to track a planet's rotation. Drifting clouds cannot provide an accurate length of day. Instead, researchers use the magnetic field of a planet.

This mosaic and other images earned Carolyn Porco the prestigious Lennart Nilsson Award (NASA/JPL/Space Science Institute)

Earth's magnetic field is tied directly to its molten core, which acts as a giant bar magnet. Magnetic field lines move out from the center in predictable patterns. The axis of the bar magnet's fields is slightly offset from the true polar axis, says Kevin Baines. "On Earth, the magnetic north pole goes into the Hudson Bay. If you sat up in space and watched Earth rotate, you would see this lighthouse effect of undulations in your magnetic field. The way we do rotation rates of Jupiter, Uranus, and Neptune is we use the fact that the magnetic field is off-center, and you watch it wobble up and down in a regular cadence as the planet rotates. In this way you can tell what the true rotation rate is."

Cassini captured complex structure in Saturn's rings (above) and erupting volcanoes on the tiny moon Enceladus (below), both of which contribute to confusion about Saturn's magnetosphere and, consequently, the length of its day (NASA/JPL/Space Science Institute)

The problem with Saturn is that the magnetic field is exactly aligned to the pole. It is offset by less than a tenth of a degree. Both Voyager spacecraft seemed to detect a faint wobble in Saturn's magnetic field, so experts thought they could determine Saturn's spin rate, which would allow them to track wind velocities and cloud structure. But it now appears they were wrong, says Baines. "Saturn has another problem as we're discovering with Cassini: there's a lot of stuff happening in the environment around Saturn. You've got Enceladus spewing out geysers of water. That stuff gets ionized and eventually slams into Saturn's atmosphere. It loads up the magnetic field lines and distorts things. Voyager may have been seeing the effect of that rather than the internal rate." Cassini's Carolyn Porco adds, "The periodicity in the radiation, called the Saturn

Kilometric Radiation, is what people were taking to be the Saturn rotation rate until Cassini got there. Now we see that that's changing. It probably has to do with mass loading on magnetic field lines. We're seeing the ionosphere rather than the actual magnetic field." With the new and improved Cassini data, astrophysicists are proposing three rotation rates. The estimates vary in length by up to half an hour. Says Baines, "We still do not know the length of a day on Saturn, even though we're staring right at it."

WINDS

A fundamental of meteorological studies is wind speed. Without the value of Saturn's daily rotation, winds cannot be understood with certainty. But scientists are beginning to discern a pattern of winds throughout the outer Solar System (see the wind diagram in Chap. 7). Winds are better understood for Jupiter, Uranus, and Neptune, as each planet has an offset magnetosphere. Saturn's wind speeds are understood in a relative sense. Jet streams blow at different rates in relation to each other. Currently, winds on Saturn are still documented using the Voyager data for Saturn's length of day. These yield absolute wind speeds higher than any other world. But without knowledge of Saturn's true spin rate, are those speeds valid? Does the Ringed Planet have the strongest winds of all planets?

Kevin Baines has been studying the problem of winds on the outer planets for some time. "If you believe in the Voyager interpretation then yes, it does, barely. It barely beats out Neptune. But it may be that what we think is prograde isn't really prograde.[1] If you tried to make Saturn look like Jupiter, all you'd have to do is move the zero line over to about 40 m/s. So if we say that what we thought was 40 m/s is really zero, now your diagram looks a lot like Jupiter, with more or less equal prograde and retrograde winds." A shift in how we chart Saturn's winds makes sense to Caltech's Ingersoll, too. "Wouldn't it be nicer if Saturn's rotation period split the difference between the atmospheric period (of some jets going eastward and some going westward). It would make Saturn look a lot more like Jupiter if the internal rate were somewhere in the middle." If analysts change their estimates of the length of Saturn's rotation by just 3 min, the zero wind speed lines move up or down the scale by 40 m/s. This is why scientists feel it is so important to understand what Saturn's true spin rate is.

Complicating the issue is the superrotation of Saturn's equatorial region. As on Venus, masses of air race around the planet independent of its rotation. A vast central river of air, whose boundaries lie at roughly plus and minus 10° from the equator, rushes eastward at huge velocities. It appears that belts and zones on Jupiter and Saturn alternate in direction, and that most winds blow toward the east. But the superrotating belt of moving air skews our view of the planets' wind directions, says JPL's Baines. "If you took that central equator region out and just looked at the other latitudes, Jupiter looks about even. You've got flows going eastward and

1. See Chap. 1: A New Spin on Things.

flows going westward. But on Saturn, everything really is pretty much prograde. There is very little that goes to zero or retrograde."

What intrigues Saturn specialists are the regions of retrograde winds. Each of those west-blowing jets generates bizarre storm systems. For example, at the center of one of those retrograde jets sprawls a line of features called the String of Pearls, a 60,000-km-long line of clearings in Saturn's upper cloud deck. Strange, donut-shaped clouds occupy another retrograde jet. The clouds resemble smoke rings, with clear air in the center. Some of the rings have lasted for several years.

Baines has a theory to explain the bizarre formations at these sites. "These retrograde jets may line up at these places so that it's a zero velocity site, and maybe when you get an updraft there it's able to just keep going up to where you can see it. It's a sort of chimney effect, in that the local air is still and allows uninterrupted convection. In other places these features might get disrupted."

Saturn's "String of Pearls," a 37,000-mile-long chain of clearings, is seen in this infrared view. Bright areas indicate openings in the clouds where heat flows from the interior (NASA/JPL/LPL)

STORMS ARE BREWING

Pearls and donuts are just the beginning of the meteorological theatrics Saturn has to offer. In the southern hemisphere, smack in the middle of one of those retrograde wind streams, Cassini discovered Thunderstorm Alley. Like the crackle of an AM radio station in a lightning storm, Cassini detected powerful activity using its radio ears. The massive storms last for days, weeks, and even months. To underscore the unsettled nature of Saturn's weather, when the Voyagers visited Saturn in the 1980s, there was no storm alley in the southern hemisphere. Some features in the northern hemisphere looked like they might be convective clouds, but no thunderstorm-like convection was visible south of the equator. With Cassini's arrival a little over a decade later, the southern hemisphere appears far more active.

Thunderstorm Alley's clouds are very dense, and darker in the infrared than any other clouds on Saturn. Saturn is a big place with a lot of weather; the planet's geographically confined lightning begs a question. "Why there?" Anthony Delgenio of NASA's Goddard Spaceflight Center asks. "No one knows. The depth of the convecting clouds – the roots of those clouds where most of the lightning is probably being generated on the Jovian planets – is deeper down on Saturn than it is on Jupiter, and there's lots more obscuring clouds and haze above it. So it's a lot more difficult for lightning flashes to escape, if you will, to where we can see them. There's too much stuff getting in the way. So there may be a lot more lightning happening on Saturn than we can observe, but it may be too deep for us to notice."

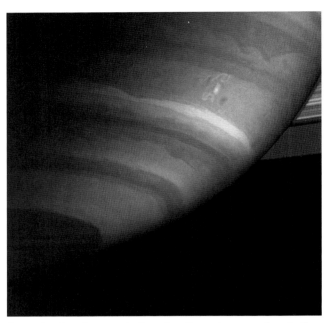

A massive thunderstorm boils up into Saturn's stratosphere in the region called Thunderstorm Alley. The complex Dragon Storm is the size of the continental United States and is associated with powerful radio bursts linked to lightning (NASA/JPL/Space Science Institute)

Researchers believe the clouds are manifestations of what's going on down below. Lightning activity appears to be associated with water clouds, as it is on Earth, Delgenio says. "From what we know about the composition of the atmospheres of Jupiter and Saturn, water is the only condensable constituent that is sufficiently abundant to create the kind of vigorous convection that would lead to lightning." The water clouds on Saturn are buried deep; water first condenses around twenty bars of pressure about 200 km down from the visible cloud tops. Internal heat triggers convection of the atmosphere, bringing the water up to about 10 bars, a vertical trip of about 70 km. At that altitude, it turns to ice.

It is widely believed that lightning on Earth occurs at the level where liquid water is turning to ice. Particles of ice and liquid water collide, causing a separation of charge. Modelers expect lightning on Saturn to begin at about the 10-bar level, about 100 km down from the dark clouds seen in visible light. Upwelling drags clouds up for 100 km, carrying substances from the depths that have been "cooked" at depth, says Kevin Baines. "We've looked at lightning chemistry; we've gone back and looked at the Miller/Urey experiment. If you zap methane long enough, like in the Miller/Urey experiment, you'll get dark stuff. We're thinking that the lightning on Saturn, which is always at the 30° latitude, is zapping methane-laden atmosphere so that what we're seeing is severely processed carbons. In fact, the only thing we can see that matches the spectrum of those dark clouds is soot." Similar processes are not evident on Jupiter. The reason may stem from Saturn's lighter gravity. There, where particles are forming at the level of 10 bars, smaller particles can live a lot longer without being pulled down. The same type of particles on Jupiter form at 5 bars, so the atmosphere can't hold it up like it does on Saturn. Saturn also has twice the abundance of methane as Jupiter, affording more methane to "cook" into dark hydrocarbons.

Saturn's brightest clouds, those of ammonia crystals, form at the tops of the thunderheads. The ammonia signature in Cassini's instruments appears directly above the lightning-forming structures. As on Jupiter, ammonia clouds dissipate within days to a week, so the thunderstorms are active at the time of the observations, developing in real time. The ammonia clouds are, essentially, an uplifted bubble of air. Ammonia is abundant enough to condense at the 1.5 bar depth, but temperatures are too high. As thunderstorms drive a piston of air up from depth, the ammonia gas erupts into higher air and condenses into distinctive clouds. About a day later, dark material comes up after it. The dark clouds persist for weeks or even months, and can be thousands of miles across.

Thunderstorm Alley may not be the unique place that current observations imply, cautions Delgenio. Rather, the regional lightning activity may be a function of the nature of gas giants. "We're familiar with Earth's atmosphere evolving chaotically on timescales of days to weeks. The problem with the Jovian planets is that they're so much bigger that they probably exhibit some of the same aspects of unpredictable behavior, but those may evolve on much longer timescales. So we see something going on at a particular latitude in the southern hemisphere and we wonder what's special about that latitude. The answer is probably that there is nothing special about that latitude except that right now Saturn is in a configuration where that is the place where moisture gathers or converges and provides conditions that are conducive to forming thunderstorms. If you came back 10 or 15 years later, the same thing might be happening somewhere else on the planet. There is nothing on Saturn that would select one latitude over another to form these storms, unlike Earth, where we have a surface with continents and oceans."

Saturn's polar regions also challenge traditional atmospheric models. One of its most striking mysteries blankets the northern pole. There, where ring shadows have darkened the surface for years, clouds are distinctly bluer than the rest of the planet. Andy Ingersoll points to Saturn's unique ring system as part of the reason. "Because the rings are there shading one hemisphere or the other, you have enormous seasonal effects on Saturn. When Cassini got there, most of the northern hemisphere had been in shadow for nearly 10 years, and yet through the gaps in the rings certain latitudes were illuminated, and they looked blue. All the classic pictures from 2004/2005 show this nice blue hemisphere peeping through the gaps in the rings, and the southern hemisphere is that yellow-orange smoggy color. Now that we're going through equinox and the hemispheres are switching roles, it looks as if the northern hemisphere is turning smog-colored."

Thunderstorm Alley gets its name from storms like this one, which is stirring up the cloud deck as it moves eastward. Inset: Cassini snapped this image of Saturn lightning in dim light (NASA/ JPL/Space Science Institute)

Even as the shadows of Saturn's rings shrink away from the northern hemisphere at equinox, the north retains its shift in cloud color toward the blue. The southern hemisphere is golden, but as winter returns to the south, scientists believe the colors will reverse (NASA/JPL/Space Science Institute)

"We still don't know what causes the northern blue," says Cassini's Carolyn Porco. "It's beautiful, isn't it? What a treat! What a present that it turned out to be so beautiful. We think it has to do with the fact that it's the winter hemisphere; we got there at the depths of northern winter. The rings cast a shadow that really makes the atmosphere cold. One idea is that perhaps the clouds just sink because the air gets so cold that the level where clouds can form gets lower and lower in the atmosphere. Above the clouds gets clearer and clearer, and you're just getting a lot of Rayleigh." As Saturn has passed equinox, and the northern pole has warmed, the blue tint is fading. Observers are watching for it to appear in the south as the ring shadows shift to that hemisphere.

The poles of Saturn display other wondrous features. In the south, a vast whirlpool gazes from concentric cloud bands like a Cyclops. The storm's

The moon Mimas soars above northern cloud tops in this true color view of the northern winter blue (NASA/ JPL/Space Science Institute)

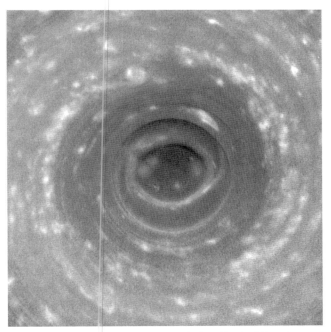

Saturn's south polar vortex is a vast whirlpool locked directly over the south pole (NASA/JPL/Space Science Institute)

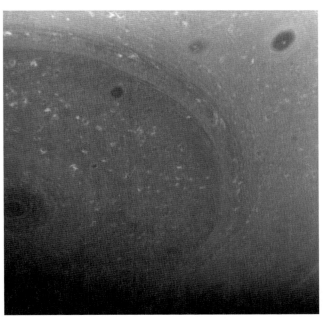

Weather in the vicinity of the south polar vortex (NASA/JPL/Space Science Institute)

cliff-like rim rises 25–40 miles high. Clear air affords views nearly twice as deep as the planet's visible cloud deck. Despite 350 mph winds, the storm remains locked directly over the south pole.

Across the northern hemisphere, a colossal hexagon drapes over territory the diameter of two Earths. Glimpsed by the Voyagers in the 1980s, the

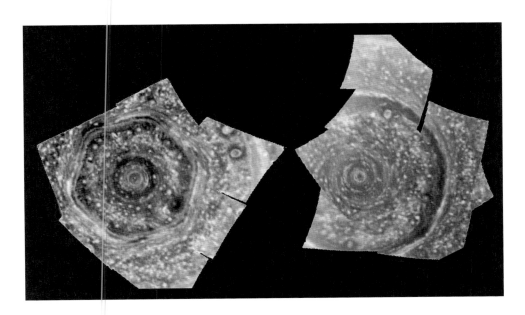

Comparison of Saturn's north (right) and south polar regions (NASA/JPL/Space Science Institute)

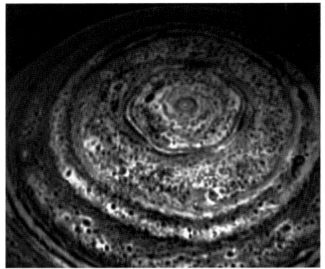

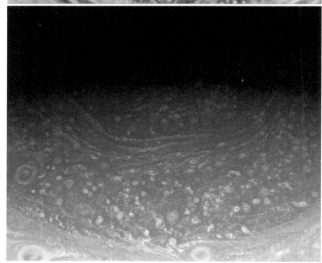

Three views of Saturn's mysterious hexagon. Top: VIMS infrared view with clouds silhouetted against the heat of the planet, taken while the hexagon was still in the darkness of winter night. Center: the same infrared view reversed to show the approximate appearance of clouds in visible light. Bottom: ISS image of the hexagon emerging from seasonal darkness shows detail in higher resolution (NASA/JPL)

baffling stream of air is stable and long-lived. To some, the Voyager images suggested that the hexagon might be a disturbance that was being forced by some adjacent vortices or spots. A large oval vortex could be seen near its border, and it was thought that the disturbance might be related to it. But Cassini images reveal that the hexagon is still there, and it still looks about the same as it used to in terms of size and shape. However, the nearby vortex has died out. An Oxford team led by Peter Reed has generated hexagon-shaped formations in a laboratory tank. The dye-filled tank is rotated, and then a shear is run across it. The hexagon's cause is not well understood, making it one of the great planetary mysteries today (see Box "The Hexagon").

Saturn's internal heat undoubtedly contributes to its variety of cloud forms. The source of internal heat on the gas giants is poorly understood, but on Jupiter it is thought to be the remnants of gravitational collapse. The planet is still condensing. Voyager data indicated that there was very little helium on Saturn. Researchers thought the helium had separated out into droplets, raining down into the center of the planet. The helium drizzle, shifting the planet's mass toward the center, would create heat. But the most recent results from Cassini suggest that there is more helium than originally thought. Saturn's heat source remains a mystery.

TO EVERYTHING, A SEASON

Seasons have a dramatic effect on Earth's weather. Hurricanes have a season, as do monsoons. Saturn, in its leisurely 29.7-year circuit around

The Miller/Urey Experiment

The origin of life stands today as one of the greatest scientific mysteries. In an effort to explain it, Russian biochemist Aleksandr Oparin wrote a pamphlet describing how amino acids and other building blocks of life might be generated by energy in a reducing atmosphere similar to those of the gas giants. His work was published in 1922 but did not reach the west until the 1930s. By then, British geneticist J.B.S. Haldane had come to similar conclusions. But the laboratory proof of their work would not come until 1953, when Harold Urey and his graduate student Stanley Miller set up an experiment. Using a sealed glass chamber filled with methane, ammonia, hydrogen, and water vapor to simulate conditions thought to exist on the primordial Earth, Miller and Urey sent electricity through the gaseous brew. Within a day, they discovered a thin surface layer of hydrocarbons floating on the water at the bottom of the chamber. After a week of operation, a brown scum had formed throughout the chamber. This dark material contained several amino acids, including alanine and glycine. The dark organic materials have counterparts common in the outer Solar System, especially on Titan. These organics are known as tholins.

The Hexagon

When the Voyagers encountered Saturn in the early 1980s, their courses took them through the Saturnian system in roughly equatorial passes. This trajectory enabled the craft to fly close to many of Saturn's moons, essentially unknown worlds at the time. While the Voyagers charted the rings, observed the atmospheres of Saturn and Titan, and mapped many of the moons, they departed with secrets still hidden in their data.

Few Voyager images revealed the polar regions, and all were taken at extremely oblique angles. Details were difficult to see, but when British scientist David Godfrey and others stretched the images to simulate a polar view, the baffling hexagon of flowing clouds appeared. They would have to wait for over two decades for the Cassini/Huygens mission to get a better look.

Cassini arrived in the throes of Saturn's northern winter, with the pole in darkness. But Kevin Baines had an experiment aboard that was oblivious to the night. "I don't care about darkness, because my instrument, VIMS (Visual Infrared Mapping Spectrometer) sees the heat of Saturn. So we use Saturn as its own indigenous light source. You just watch that radiation coming up through the atmosphere; where the clouds are, it's black. You're watching the clouds in silhouette." Researchers took the VIMS images and inverted them, so that dark clouds became light and the bright background became dark, simulating what the naked eye would see. Says Baines, "Voila, you've got a beautiful picture of the hexagon."

Cassini's Imaging Science System (ISS) has 80 times the resolution of the VIMS instrument. As the north pole drifted into light with the onset of northern spring, ISS began to see details as never before. The Hexagon covers an area wider than two Earths, surrounding Saturn's north pole at 77° latitude. Winds roar along the hexagon at 100 m/s (220 miles per hour), splitting the feature's edges into swirls, spirals, and streamers. The formation is essentially a trough, bordered on either side by a wall of clouds rising in steps.

ISS was a perfect complement to VIMS, which could gaze more deeply into the atmosphere. The VIMS data only increased the mystery of the atmospheric structure. Images revealed the hexagon in the same shape as was seen 30 years ago, still buried deep in the atmosphere on the order of 50–100 km underneath, where it appears in reflected sunlight [by the Voyagers and ISS]. The formation still seems to rotate at the rotation rate of Saturn itself while individual clouds move around it like a racetrack at very high speeds. The latest Cassini data shows that the hexagon has moved less than half a degree in 3 years.

This feature, unique in all the Solar System, mystifies researchers. With the permanence and stability of the hexagonal structure, atmospheric dynamicists will be entertained for years to come.

the Sun, endures seasons that last nearly 7½ years. At 26.75°, its axial tilt is similar to Earth's (23.4°), so its seasons vary about as much. Storm activity from the Voyager encounters to the Cassini mission seems to have changed

Three reminders of how different Saturn's elegant weather is from that of Jupiter (NASA/JPL/Space Science Institute)

significantly in terms of global patterns. Is this due to a seasonal effect? Anthony Delgenio says, "It's certainly not out of the question that as you change the seasons you could get responses from storms. The problem with that is that in order for seasons to affect the weather, the solar heat has to be at altitudes low enough to heat the atmosphere."

On Saturn, the sunlight doesn't penetrate very deeply because of its photochemical hazes and high-altitude clouds. Thunderstorms spawn at low levels, far beneath the visible cloud deck. The storms are driven by temperature gradients. They occur when the air is warmer at the bottom than at the top. This is the reason that thunderstorms on Earth are most frequent in late afternoon; the sunlight has heated the ground surface. But Delgenio points out that thunderstorms can be generated equally well if temperatures above are cooled rather than temperatures below rising. "So it's possible that the seasonal cycle modulates how cool it is at the top, and that has some effect on thunderstorm activity. But it's hard to understand how it would have enough of an effect to produce major changes in a phenomenon. That having been said, we know that in the equatorial region of Saturn there are these storms that pop up sometimes. For example, Hubble saw this huge equatorial storm back in the 1990s. There does, in fact, seem to be some modulation with the seasonal cycle. People are still scratching their heads trying to figure out how these seasonal effects could impact the [deeper] thunderstorms rather than just the upper level clouds and hazes. Nonetheless, there is some evidence that these things come and go with a little bit of a mind toward what the seasons are doing. It is scant, it's very anecdotal, but it's there."

Scientists such as Delgenio still have to figure out whether or not the seasons have a more significant effect than current estimates indicate. One issue they must grapple with is the long timeline in the outer Solar System. Not only do the outer planets have long years, they have a lot of mass. As Delgenio warns, "One of the things that you have to be a little circumspect about when you're studying the Jovian planets is that because they are so big and so massive, a lot about them changes very slowly."

TITAN

If a cosmic catastrophe plucked Titan from its orbit around Saturn and cast it into its own independent path, many scientists would argue that it was the most important planet to study in the entire Solar System. On the cosmic yardstick, Titan measures nearly the size of the planet Mercury. Its 5,150-km (3,200-mile) diameter makes the moon second only in size to Jupiter's Ganymede. Titan has an atmosphere denser than Earth's, and its weather will be among the most alien we visit on this journey. Winds move in great atmospheric tidal waves across the face of the moon, pulling sand dunes into parallel lines. Titan has the closest thing to Earth's

Rossby Waves and the Hexagon

Kevin Baines relates the Rossby restoring force to Saturn's hexagon this way: "You have all these east-west moving pieces of air. Say you place a rock in the middle of one of these jets. The stream has to go around to the north, but it has the memory of where it wants to be. It wants to be back where it was. It will start wandering south again, but then it overshoots and wants to go north and then south again. It undulates in a sinusoidal line. You can create a six-sided wave feature going around the planet. If you look at that feature from over the pole, it looks like it has straight lines. Anywhere along that 'straight' line it's changing its latitude, climbing north, and then reaching back down south until it rounds the corner, but that corner is also part of that same sinusoidal feature."

Titan's Hotei Arcus region may host cryovolcanic activity. Super-chilled geysers steam from frozen hills as distant methane rains feed dry riverbeds (© Michael Carroll)

Titan is the penultimate moon in size, nearly the diameter of Mercury (© Michael Carroll)

hydrological cycle that we have seen on any solid-surfaced world, and its complex chemistry promises to be a playground for planetary scientists for centuries to come.

Even through the telescope, observers could tell something exciting was going on at Saturn's giant moon. Unlike any other moon in the Solar System, Titan had atmosphere, and a lot of it. Methane left a distinctive signature in the spectrum coming through their telescopes. Dark lines crossed the moon's ruddy light, betraying the pattern of methane's unique fingerprint. Clearly, the gas was absorbing some of the light reflecting back to Earth. The question was, *how much* atmosphere was there? The nitrogen in Earth's atmosphere makes it difficult to "see" nitrogen on other worlds. If Titan had a lot of nitrogen mixed with the methane, it might have enough pressure to enable that methane to liquefy, perhaps even into lakes or planet-wide oceans. Scientists had been postulating liquid methane or ethane on Titan's surface since the late 1960s. The chemistry and temperature seemed just right for methane's triple point, the point at which methane can exist as a vapor, liquid, or solid (in the form of ice). Our Earth is at the triple point of water. It is water that dominates our surface erosion. Could methane dominate the landscape of Saturn's moon?

The atmosphere was equally peculiar. The first close look came through the robot eyes of Pioneer 11 in 1979. Scientist Peter Smith was on the team. Among the most puzzling things Smith saw was the super-high polarization of Titan's atmosphere. What kind of particles could be so strongly polarized and behave as those in Titan's air? Smith's team was bewildered. "It was very strange, and we spent a lot of time making models to explain what could be causing that polarization. The other thing that was mysterious was that the light was forward-scattering, so the particles had to be large to scatter light forward, and yet they had to be small to produce high polarizations. How can you have particles that are both big and small at the same time? It wasn't until the late eighties that Bob West and I figured it out. That is that there are clusters of tiny particles that make big particles. They are loosely bound together. The smallness of the particles produces the polarization and the largeness of the cross-section explains the forward-scattering."

In 1981, mission planners targeted Voyager 1 and 2 to fly by the moon in an attempt to unravel more of its mysteries. Voyager 1's flyby brought the craft within 4,000 miles of the surface. Mission planners hoped to get glimpses, at high resolution, of Titan's surface. In fact, Voyager's high-resolution camera would be able to make out objects only a few kilometers across. But an opaque, orange fog stymied Voyager's robot eyes, not even surrendering small breaks in the clouds for the frustrated imaging team (see Chap. 3). The probes revealed a dense atmosphere, perhaps dense enough to cause those methane rainstorms that had so captivated past authors and writers. Among the stories penned, British author Sir Arthur Clarke published a 1971 short story called "Imperial Earth" in which he described the "methane monsoons" of Titan. But to see his proposed downpours and their

The Voyagers were met with an impenetrable wall of fog during their Titan encounters (Voyager 1 photo courtesy NASA/JPL)

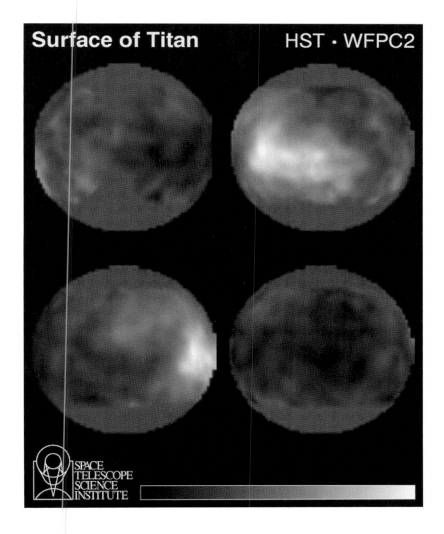

Surface of Titan   HST · WFPC2

SPACE TELESCOPE SCIENCE INSTITUTE

Four near-infrared views of Titan taken by the Hubble Space Telescope. Were those surface variations continents and oceans of hydrocarbons? (Space Telescope Science Institute)

Cassini's VIMS and ISS instruments were able to peer through Titan's fog, discerning surface features (NASA/JPL/ Space Science Institute)

possible oceans below, we would need new instruments.

Ground-based observers made an important discovery after the Voyagers: although Titan's haze was opaque at visible wavelengths, astronomers were able to peer through it in the near-infrared. Studies carried out with the Hubble Space Telescope and several ground-based observatories revealed variations on Titan's surface. Was it possible that those blurred features were continents and oceans? The ultimate answer would have to come from close-in studies.

NASA and ESA combined forces to launch the school-bus-sized Cassini spacecraft in the fall of 1997. Settling into orbit around Saturn in July of 2004, the Cassini–Huygens spacecraft has revolutionized our knowledge of the cloaked moon. Cassini's Visual and Infrared Mapping Spectrometer (VIMS) and its Imaging Science Subsystem (ISS) are sensitive to the near-infrared light that seeps through Titan's foggy air. The experiments are able to image the surface, just as Hubble had two decades before. In addition, the Cassini orbiter's synthetic aperture radar (SAR) returns fine detail from the surface. With each loop around Saturn, Cassini passes by the moon, turning its radar toward the surface to map a thin strip tens of miles wide and hundreds of miles long.

The first images were mystifying. Strange cones of bright areas fanned out across dark plains. An arc of dark material reminded some analysts of the edge of a mountain, perhaps something like Venus's famous pancake domes. Several regions were swathed by "cat scratches," long, sinuous

The first radar image of Titan's surface initially evaded scientific interpretation (NASA/JPL)

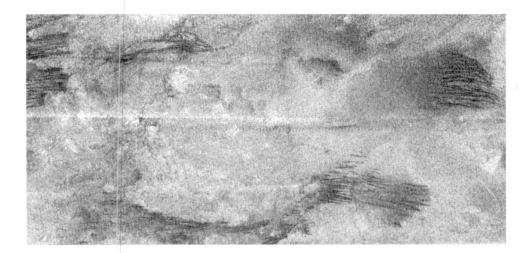

Titan's "cat scratches" (NASA/ JPL)

parallel lines. What the early passes clearly did not show were lakes or oceans, says Jonathan Lunine of the University of Arizona's Lunar and Planetary Laboratory (LPL). "We've swung back and forth with Titan. After Voyager, the idea of a global ocean of liquid methane was very appealing. Then the remote sensing data and the early Cassini data seemed to suggest that no, it looked rather dry." That view morphed again in 2007 as Cassini began polar passes, revealing open bodies of liquid ethane and methane. Titan is, in the equatorial regions, a sort of Arrakis, a dune world washed by rare cryogenic cloudbursts. In the polar regions, Cassini revealed the missing lakes, some the size of respectable seas. The polar regions present a world where it may rain seasonally, and where methane is a prime ingredient for the climate, both in liquid and gaseous form. As with Mars, the surface informs us about the atmosphere.

The southern hemisphere formation Ontario Lacus was the first true methane lake discovered on Titan (NASA/ JPL/Space Science Institute)

MYSTICAL METHANE

Titan's brooding skies continue to captivate. Researchers are baffled by just how much methane exists in its atmosphere. Sunlight works to break methane into other constituents. In addition to ethane, butane, and complex hydrocarbons, Father's Day barbecues would be well supplied with a drizzle of "natural gas" fresh from the orange fogs above. The rain precipitates out of methane-nitrogen clouds at an altitude of 12.5 miles (20 km). Above that cloud deck, a gap of clear air underlies a high altitude layer of methane ice crystals. The methane humidity increases closer to the poles, where methane/

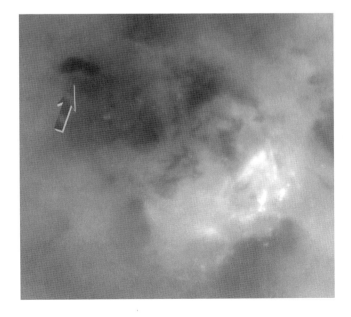

The "lake district" of Titan
displays a variety of forms,
including embayments,
peninsulas, and drainage
systems (NASA/JPL)

ethane lakes have been found. Scientists estimate that the current levels of
methane should disappear in a few million years. Something is replenishing
the gas, says Goddard Space Flight Center's Anthony Delgenio. "The canon-
ical story for Titan is that the methane molecule is destroyed by ultraviolet
radiation, and that's how those complex hydrocarbons make the haze in the
stratosphere. The idea is that the process is fast enough so that if that were
the only process going on in Titan's atmosphere, all the methane would
have disappeared in something like 10 million years. But since we still see a
few percent's worth of methane in Titan's atmosphere, that means there
must be a source of methane replenishing the atmosphere, and that source
must be the surface. That source could be from locations of open liquid
methane on the surface, which we appear to have as lakes in both the northern and southern polar regions. We do now have good evidence that there
are, at least in the polar regions, sources of liquid methane. There are a
number of people who think that cryovolcanism may also be an important
factor."

Jonathan Lunine is part of a team that has been studying Titan's evolution in an attempt to discern why Titan has so much methane today. The
team believes that Titan's interior pumped methane into its skies during
three developmental epochs. In its formative years, as the moon accreted
from the solar nebula, a rocky core formed beneath a water mantle. A water-
ice crust topped the mantle. During its first several-hundred-million years,
heat from the moon's formation combined with the warmth of radioactive
elements in the core to melt through the crust, releasing methane.

The second release probably occurred about 2 billion years ago, when
Titan's silicate core began to convect. This geological burst of heat again

melted the crust, causing methane outgassing. Ammonia mixed with the water-ice would have helped to serve as a natural antifreeze.

Lunine explained what happened in the most recent epoch, which has left Titan with a methane-rich atmosphere. "In that last stage, somewhere between half a billion and a billion years ago, the cooling of Titan gets to the point where a layer of [water-] ice forms immediately beneath what has been a methane/clathrate crust [methane trapped within a lattice of water-ice]. The cooling crust thickens, and convection begins in the ice crust itself. Upwelling plumes of solid material force the release of methane from the clathrate above it. We argue that thermal conditions changed so that, plausibly, you might have geysers or places where methane is leaking out of the ground." The most recent epoch may have opened with a catastrophic volcanic tirade whose eruptions resurfaced the entire globe. Isotopes – identical gases with a different number of neutrons – in Titan's atmosphere reinforce this scenario of alternating quiet and violent geologic episodes.

METHANE MONSOONS?

Wherever it is coming from, Titan's methane plays a critical role in the weather. Like water on Earth, methane cycles regularly, raining out of the sky, pooling as surface and subsurface liquid, then evaporating as vapor to return to the sky once again. Before Cassini charted the humid northern and southern latitudes, its orbits carried it across the desiccated equatorial regions. Vast seas of dunes and dry mountains spread before its radar eyes (the Belet Sand Sea alone is over 3,000 km across). But the radar also saw the river valleys, eroded slopes, and alluvial fans marking outflow regions. Where had all the rain gone? The dry equator may simply be evidence of the currents in Titan's atmosphere. Like Earth, Titan has Hadley circulation (see Chap. 1). On Earth, heated air rises over the equator, migrates toward the poles, and is stretched into spirals as Earth turns under it. But Titan turns at a leisurely pace, once every 15 Earth days. Its simplified Hadley circulation resembles that of Venus, with air rising over the equator and looping toward the poles as it descends. The result – the equatorial region dries out, while moisture migrates to the polar regions.

As scientists began to understand methane's poleward migration, NASA's Infrared Telescope Facility atop Hawaii's Mauna Kea caught a gigantic tropical storm in the act. The equatorial storm grew and drifted to the southeast. Cassini later imaged several methane storms in various stages of formation. These were lucky events, explains Tony Delgenio. "The way Cassini observes Titan is just every once in a while. The Cassini mission is designed to observe the entire Saturn system. Any given instrument at any given time is observing the many objects in the Saturn system, whether it is Saturn itself or the rings or one of the many other satellites or the magnetosphere. The only time that we're observing Titan is when we come close enough for a Titan flyby. That happens no more than every 16 days, and often times a lot less frequently."

Although Titan has clouds and appears to have methane rainstorms, those storms are a lot less frequent than meteorologists had hoped, Delgenio notes. "Unfortunately, the relative humidity of methane in Titan's troposphere near the surface is just not high enough for clouds and rainstorms to break out most of the time. It turns out that it only gets that way at selected locations every once in a while. We can go an entire flyby of Titan and not see a single cloud."

The methane concentrations may remain constant globally but may change regionally with the seasons as the pattern of rainfall changes. The precipitation events themselves may be sporadic, even though the global rate of evaporation is slow and steady. Titan's weather may resemble the most severe Earthly drought conditions for centuries or millennia, broken by brief periods of flash flooding. Delgenio likens the situation to a slowly accruing bank account. "It's like saving money in a Christmas Club every paycheck throughout the year, and then in December you take it all out and spend it all at once."

Patterns of methane clouds and precipitation are difficult to chart. On Earth, just 93 million miles from the Sun, prodigious amounts of energy enter our climate system constantly. This makes Earth's atmosphere a vigorous and active place. The solid surface beneath the atmosphere breaks the flow of air and disrupts temperature gradients, so that the environment cannot maintain weather systems for extended periods. Our capricious meteorology stands in stark contrast to that of distant Titan, Delgenio stresses. "If you were to take any day's weather satellite image of a region of Earth, you could see examples of just about all the things that are important to Earth's meteorology. You'd see numerous thunderstorms, cold fronts, and warm fronts, evidence of the Hadley cells. On Titan it doesn't work like that. There is much less sunlight entering the atmosphere, which means that there is less need for these individual weather events. Rainstorms play a role in Earth's atmosphere by taking warm air near the surface and bringing it up to higher altitudes. There's a lot less need for that on Titan because there's so much less sunlight entering that atmosphere to begin with. As a result, an occasional storm may be all that is needed."

Analysts have likened atmospheric heating to a pot of water on

The methane cycle on Titan may be quite similar to the water cycle on Earth (© Michael Carroll)

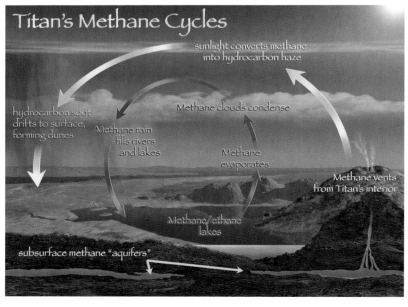

Titan's Methane Cycles

sunlight converts methane into hydrocarbon haze

hydrocarbon soot drifts to surface, forming dunes

Methane clouds condense

Methane rain fills rivers and lakes

Methane evaporates

Methane vents from Titan's interior

Methane/ethane lakes

subsurface methane "aquifers"

a stove. When the flame is first lit, an occasional blob of water will rise from the bottom of the pan to the top, making a little turbulence. As time goes on and the water gets hotter, more and more of those blobs will rise through the liquid. The water in the pot can be very turbulent even before the water itself begins to boil. Titan's atmosphere may be similar to the beginning stage, where blobs of warm fluid come up only rarely. Earth's atmosphere is similar to the pot's liquid at the boiling stage. The rarity of Titan's dramatic weather, coupled with the fact that Cassini observes Titan infrequently for short periods, makes for challenges in discovering significant weather events, Delgenio says. "You have to watch for a long time before you even see a few examples of what you want to see. It takes a long time for things to change on Titan, and it takes a long time for us to detect changes on Titan."

Nevertheless, when Cassini arrived at the Saturn system in 2004, researchers did see evidence of convective clouds in the south almost immediately. A year later, features in the same region appeared to have changed. In particular, it looked as though the lake pattern in the southern region near Ontario Lacus had changed in several places. Observers suspect that storms may have produced new lakes in that area.

It is conceivable that as the seasons change and the convective storms shift from primarily in the southern hemisphere to primarily in the north, lakes may begin popping up in new locales. Another unanswered question concerns how those lakes will move across the planet. Will they shift from the southern hemisphere to the northern, or will they migrate across the equator? Scientists like Delgenio want to understand whether the distribution of lakes on the surface is a dynamic or a permanent characteristic of Titan's gloomy landscape. Drainage patterns certainly suggest that there had to be storms at different times in the past. Researchers believe that at least occasionally there must be strong rainstorms in the low latitudes in order to explain those geological forms. It may be that those storms are so few and far between, and the atmosphere is so dry, that the formation of permanent or semi permanent lakes in the low latitudes may not be favored. Says Caltech's Andy Ingersoll, "I think the rain is monsoonal, and my reasoning is that you see these river channels that need a very intense flood to produce them. Steady drizzle won't do it. It's got to be a pretty vigorous rain. There is not a lot of energy on Titan to evaporate the methane and then condense it to make the

Methane storm system developing over a 5-h period (NASA/JPL/Space Science Institute)

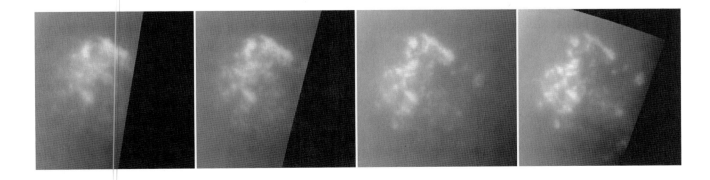

rain. It has to be intermittent and the events have to be separated in time." Titan's methane storms may be most like storms over terrestrial deserts that wet the ground but don't result in any significant buildup of bodies of liquid.

Between these rare storms, some experts believe Titan's alien skies subject the landscape to a constant drizzle. Chris McKay of NASA's Ames Research Center was studying Titan's atmosphere even before the Cassini mission. The newest data from Cassini and ESA's Huygens probe have brought a clearer picture of conditions on the strange moon, McKay says. "The rain on Titan is just a slight drizzle, but it rains all the time, day in, day out. It makes the ground wet and muddy with liquid methane. This is why the Huygens probe landed with a splat. It landed in methane mud." McKay and his colleagues estimate that the annual rainfall on Titan amounts to about 2 in. (about 5 cm). This is the equivalent to annual precipitation in Death Valley. But on Titan, that scant amount of methane rain is spread out evenly over its entire year.

ERUPTING METHANE?

Cassini's Synthetic Aperture Radar has mapped many drainage networks, river valleys, and dark basins on the surface that are either currently filled with methane or were filled until recently. These bodies undoubtedly contribute to the atmospheric methane, but another supplier of the methane may be cryovolcanism. Long before Cassini settled into orbit around Saturn, researchers had reason to believe that super-chilled cryovolcanoes might lurk

Three artist's impressions of Titan document our changing scientific views. Left: Chesley Bonestell's quintessential space art from 1944, depicting a reddish surface and thin atmosphere. Center: David Hardy's 1972 rendition shows knowledge of methane in the atmosphere, which would throw the sky color toward the blue/green. Right: View of Titan's surface, covered in ice flows and methane fjords, painted shortly after the Voyager Saturn missions (Credits, respectively, © Bonestell, LLC, © David A Hardy, 1972, and © Michael Carroll, 1983)

beneath the surface of Saturn's planet-sized moon Titan. Their suspicions were based upon three lines of circumstantial evidence. First, Titan's size and mass suggest that it harbors a great deal of internal heat, some of which is left over from its formation, and the rest generated by radioactive elements.

Secondly, Titan's orbit around Saturn is also somewhat eccentric, which may supply a muted form of the tidal friction that causes volcanism on Jupiter's moon Io. This internal heat could be expressed as some type of alien volcanic feature.

The third line of evidence is the substantial amount of methane present in the atmosphere, which must be replenished to remain so abundant.

One candidate site for erupting methane is called Hotei Arcus, an area under study by Randall Kirk at the U.S. Geological Survey. "The first look we had in radar showed the kinds of lobes and protrusions and indentations that you get with lava flows or other viscous material. Some thought these might be sedimentary deposits related to the narrow channels that were flowing into the area." Then, Kirk's team got a second radar pass, enabling them to construct stereo images of the area. The new data suggests that the channels could not have deposited the flows, which tower up to 200 m above the surrounding landscape. Kirk believes other sites display possible cryo-volcanic signatures. "A couple features seem to have a hole in the ground like a caldera (volcanic crater) with a thick, snakey flow coming out of it, similar to silicate lava flows." Another region, Tui Regio, has the same unique spectra that Hotei does and seems to be composed of similar, flow-like features.

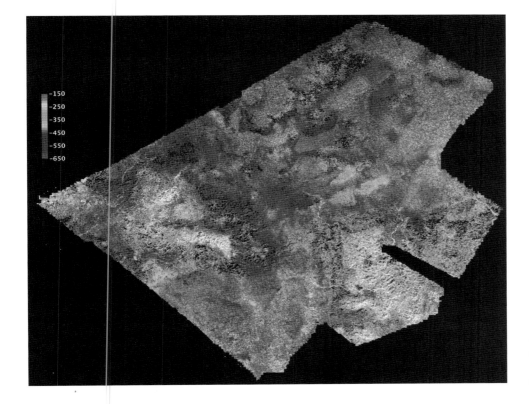

Radar image of the Hotei Arcus region, one possible cryovolcanic source (NASA/JPL/USGS)

Titan's dense atmosphere moves sluggishly. Before the Cassini mission, many researchers believed winds would be low-velocity and difficult to monitor. It was hoped that the Huygens probe would return data on wind speeds and directions as it descended through the atmosphere. But the Cassini orbiter was also able to chart the moon's winds on a global scale by using a natural weather vane – sand dunes. Cassini radar images revealed vast tracts of dunes that rival any desert on Earth. "We've been bewildered for some time by the forms we've seen in the Cassini data," says Ralph Lorenz, Cassini team member at the Lunar and Planetary Laboratory. "The exciting thing is that we can now start to make sense of things on Titan."

Lorenz and others believe that the dunes – one of the few Earthlike formations seen on the bizarre moon – may consist of pulverized water-ice. Titan's surface is made up primarily of water-ice frozen rock-hard in Titan's frigid temperatures (−290°F). Relentless winds may break down the ice into particles similar to terrestrial sand. The roughly parallel dunes are spread across regions up to 1,500 km long and are similar in form to the majority of dunes seen on Earth in sites such as the Namib desert.

But even more alien processes may be at work. A constant rain of organic material drizzles from Titan's ruddy sky. This sooty sludge may pile up into material that is blown into the dune-like formations visible in Cassini's images.

Most significant, however, is what the dunes tell us about the environment on Titan. Lorenz feels the dunes are evidence of an atmospheric phenomenon not visible anywhere else in the Solar System. The immense gravity field of nearby Saturn pulls on Titan's dense atmosphere, creating a tide of pressure that moves across its face. This tide sets up alternating east-west directional winds. The shape and location of the dunes are consistent with this unique tidal wind, which may dominate the near-surface environment. "This is probably one of the neatest implications of our study," says Lorenz. Lorenz and colleagues assembled 16,000 radar segments covering dune areas. Cassini took the radar images over the course of 4 years. With that data in hand, they were able to see that near-surface winds move primarily toward the east.

The dunes have left scientists with a mystery. Computer models of atmospheric circulation, coupled with data from Europe's Huygens probe (see later) led researchers to believe that Titan's winds would blow from east to west around the equator. But Cassini's radar maps of the dune fields clearly show that the winds must be blowing in the opposite direction. Recent work by Tetsuya Tokano may have solved this conundrum.

Tokano's research suggests that a 2-year-long "gust" of global wind forces the dunes into the orientation seen by Cassini. During most of Titan's long year, winds behave, blowing in the predicted east to west direction. But at the equinox, when solar heating shifts from one hemisphere to the other, a dramatic wind reversal causes the dunes to move in the opposite direction. On Earth, this seasonal wind reversal occurs over the Indian Ocean and other

monsoon-associated locations. The seasonal wind gusts on Titan appear to move at 1–1.8 m/s (2–4 mph). Titan's sticky organic sands begin to move as winds reach 1 m/s (2 mph), a speed higher than normal winds travel but consistent with the breezy equinox episodes.

The direction of dunes helps chart global wind direction (NASA/JPL/Space Science Institute)

HUYGENS

Cassini's VIMS, ISS, and radar systems amassed huge quantities of data that enabled cartographers to map vast areas of Titan in detail. But it was up to the European Huygens probe to give us our first direct look under the mystifying fog (see Chap. 4). The smart-car-sized probe hit Titan's atmosphere at 21,600 kmph (13,400 mph), glowing like a meteor as it burned off speed. After slowing to a leisurely 1,440 kmph (895 mph), the probe deployed a series of parachutes to slow it down. The descent was timed to coincide with a high-altitude pass by the Cassini orbiter, which served as a relay to Earth.

Huygens was designed to spin as it descended through the thick Titan fog, sampling the atmosphere, winds, and temperatures of the strange new world. Imagers on board would transmit views of the atmosphere and if the fog was thin enough, we just might see the surface.

Huygens was a spectacular success, radioing data through a 2 h 27 min descent, and continuing to transmit from the surface to Cassini for another 69 min. Huygens actually survived on the surface for a total of 3 h 14 min. What Huygens told us about Titan's weather was intriguing. A steady drizzle

Colorized version of Huygens panorama assembled from images taken at approximately 25 km altitude. The view is toward the southwest. Note the drainage features in bright highlands leading to the dark lowlands, where the probe ultimately landed (ESA/JPL/DISR images reprocessed by the author)

of organic soot tints the skies a dramatic dark red, and stains the lowlands. Much of what we know about the methane rains and atmospheric structure came from this plucky European marvel. Some of this data was discerned by stellar and radio occultations by the orbiter, but Huygens could sample gases directly and sense moisture, even in the surface. Its in situ measurements confirmed the existence of complex organic chemistry. Huygens detected a super-rotation in Titan's atmosphere. At 120 km above the surface, winds blow in the direction of Titan's spin at up to 431 kph. Winds decrease with decreasing altitude and reverse direction below 8 km. Surface winds drop to less than a meter per second. Before Huygens, models predicted a clear stratosphere from 60 km to the surface, but Huygens sensed haze all the way down.

When Huygens landed, its accelerometers recorded the probe's impact with the Titan surface. The touchdown was surprisingly gentle. One ESA commentator described the surface as having the consistency of crème brule. Apparently the ground was saturated with methane, because the heater by the mass spectrometer drove off some gases. Huygens also detected benzene, cyanogens, and ethane in the soil. Temperature and pressure sensors continued to monitor the surface conditions for over half an hour after landing. The weather report – a temperature of −179°C and a pressure of 1.47 times the density of Earth's at sea level.

Titan's rainstorms of "natural gas" and organic soot remind us of just how alien this moon's meteorology truly is. Titan joins the other gas giants as a challenge to theory and convention, and a puzzle to those who want to

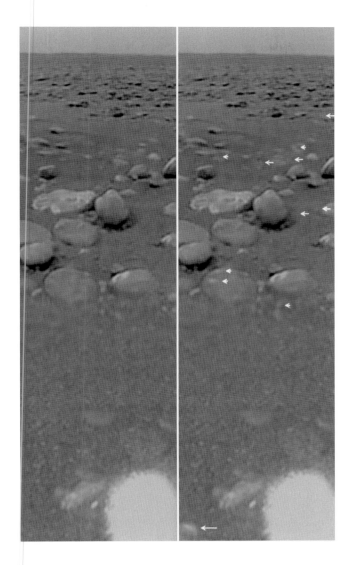

Huygens surface image shows water-ice "stones" eroded into shapes similar to river rock. The rounded rocks in the center of the image are about the size of a fist. The image on the right, taken 15 s later, shows a dewdrop that has condensed on the camera hood (lower left corner), with other possible condensation droplets marked by arrows (JPL/ESA/University of Arizona)

comprehend outer planet meteorology. As Anthony Delgenio and others struggle to understand these distant worlds, Delgenio advises, "It really helps to think about Earth in relationship to Titan. When you do, you realize how much tougher a nut the Jovian planets or anything in the outer Solar System is to crack."

Methane clouds condense over the far rim of a massive storm on Neptune. Nearly supersonic winds send them drifting through the clear air above the cyclone. Voyager 2 imaged dark storms such as this; one was wider than Earth (© Michael Carroll)

Chapter 9

Uranus, Neptune, and Triton

Beyond the orbit of Saturn lie two worlds long thought to be identical siblings. Uranus and Neptune are virtually equal in size, but the two frigid giants turn out to be quite different from each other, each unique in our planetary system.

Like Jupiter and Saturn, Uranus and Neptune have reducing atmospheres, air dominated by hydrogen, helium, and methane, similar to the makeup of the primordial Solar System. Their cores are larger, compared to each respective planet, than those of the inner two giants. Their cores have more water and are much cooler than those of Jupiter and Saturn, inspiring researchers to brand the two planets with the term "ice giants."

From Earth, prior to the Space Age, observers saw little structure in either world, assuming them to be planets chilled into quiescence. The Voyager 2 spacecraft revealed just how wrong they were. The planets both have active atmospheres, though Uranus less so than Neptune. Bands and zones similar to the inner gas giants parade across their faces. Clouds take different forms on both worlds.

URANUS

It's been called the Jolly Green Giant, the Cosmic Pea, and less charitably, "the most bland target in the outer Solar System." In visible light, its blue-green disk is featureless. Uranus's remarkable color stems from the methane in its atmosphere. Methane absorbs red light, leaving the bluer parts of the spectrum to reflect back at the observer.

Uranus has a calm, transparent atmosphere to a great depth. A ruddy haze shifts the color of the cloud deck toward the green.

Retaining a subtler version of the same belts and zones exhibited by its gas giant siblings, Uranus, as seen by Voyager 2, had its axis pointed almost directly toward the Sun, so heat was falling directly onto the pole. Evidence seems to point to the rotation of the planet as being responsible for the belts and zones rather than solar or internal energy.

Although belts, zones, streamers, and swirls are clearly seen on the other three gas giants, the weather on Uranus is far more subdued. Why?

The answer lies within the depths of these outer worlds. Jupiter, Saturn, and Neptune all put out more energy than they receive from solar heating. This internal heat drives the weather from within (see Chap. 7). But Uranus is far colder compared to its surroundings. The temperatures

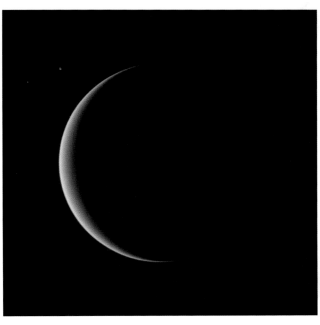

The crescent of Uranus, seen here in visible light, would appear featureless to a human explorer (NASA/JPL)

M. Carroll, *Drifting on Alien Winds: Exploring the Skies and Weather of Other Worlds*,
DOI 10.1007/978-1-4419-6917-0_9, © Springer Science+Business Media, LLC 2011

Voyager 2's image of Uranus in visible light (left) and computer enhanced (right). Note the bull's-eye pattern of belts and zones centered on the south pole, which is facing the Sun (NASA/JPL)

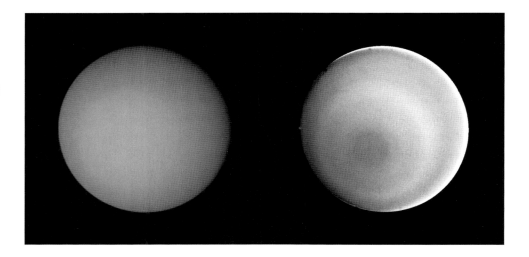

on Uranus and Neptune are nearly equal, at −323°F, even though Neptune receives only 4/9 the solar energy that Uranus does. In the case of Uranus, its temperature appears to be in equilibrium with the incoming solar energy, leading to an atmosphere that is less mixed from interior to surface.

"Talk about an energy crisis," says JPL's Kevin Baines. "We don't understand any of the outer planet heat sources!" Researcher Dave Stevenson of Caltech believes the lack of heat flow may be a temporary condition. He points out that if a massive moon hit Uranus in the distant past and tipped it over, the catastrophe would shut down the heat source for a while. For now, Uranus's cool core remains an enigma.

Many researchers attribute the low Uranus heat flow to whatever caused the planet to spin on its side. Heidi Hammel, an outer planet expert at the Massachusetts Institute of Technology, thinks a titanic collision with a roaming planet is the best explanation. "A major collision like that would so effectively perturb the atmosphere and disrupt it that any internal heat source that's there would have the opportunity to radiate far more effectively during a time of great turbulence like that. I would think that the heat source would be shut down completely." There are other possible explanations for Uranus's lack of heat flow. One has to do with convection, Hammel says. "There is an analogy to the salinity layers in Earth's ocean that prohibit convection from occurring. People have made analogous arguments that the icy mantle in Uranus is layered and prohibiting convection from occurring. That doesn't mean the heat isn't in there. It's just not getting out."

The Voyager encounter of 1986 uncovered another mystery pertaining to Uranus's heat flow: in the upper atmosphere, the north pole – hidden in darkness – was actually warmer than the day-side pole. Voyager trained its UV instruments on the star Gamma Persei as the star passed behind Uranus. During this occultation, the star set over the sunlit south pole, then exited over the dark polar region in the north. The sunlit pole was found to be 450° cooler than the night-side north pole.

CLOUDY OUTLOOK

Although methane lends color to the planet, Uranus's atmosphere is primarily hydrogen and helium in roughly the same ratio as in the Sun. A high-altitude layer of complex hydrocarbons blankets all four gas giants. Solar ultraviolet radiation transforms methane into a fog of acetylene, ethane, polyacetalene, and other organic molecules.

Voyager visited the Uranian system at a time when the south pole of the planet was pointing at the Sun. A polar hood of dark haze spread across the south pole. A bright band, concentric to the pole, forms a ring down at the 50° latitude. The band may consist of methane clouds and upwelling hazes. A cold region stretched from about 10 to 40° latitude. Dark bands bracketed the bright one at 20 and 65° latitude. These regions reveal deeper, descending air masses. Unlike Jupiter, where bright clouds are coldest, Uranus's bright band is warmer than the dark areas. The difference may result from the difference in latent heat on Uranus's methane and Jupiter's ammonia clouds.

The cloud deck on Uranus is buried deep in the atmosphere, beginning at the 0.9 bar pressure level. Above these clouds, the upper atmosphere consists of molecular hydrogen, some atomic hydrogen, helium, and haze layers of hydrocarbons. Scientists monitored Voyager's radio waves while the spacecraft passed behind the planet and were able to chart lower structures. The base of the cloud deck seems to rest at the 1.3 bar level, where temperatures are about −314°F (−192°C). Many narrow cloud bands encircle the planet near the equator.

Winds are difficult to judge. In both Voyager and Hubble Space Telescope images, few distinct clouds appear. Voyager was only able to image the southern hemisphere, as the north was in darkness. (See Chap. 3 for a description of Uranus's odd orbit.) Discrete clouds that have been followed yield equatorial wind velocities of up to 360 mph. The visible clouds are crystals of methane ice, and may be welling up from below.

The reasons for the dearth of clouds on both Uranus and Neptune compared to their gas giant

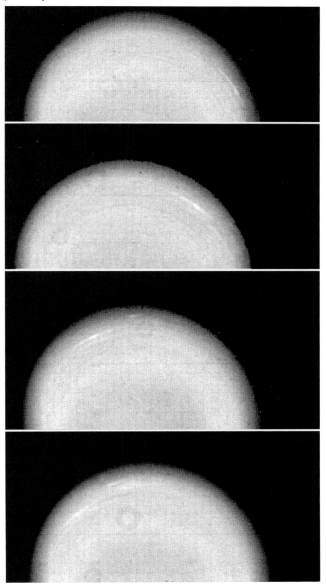

Voyager 2 took these four images of Uranus over a 4.6-h period on January 14, 1986. Exaggerated contrast brings out detail. Note the two long clouds that move counterclockwise with the rotation of the planet. The clouds turn at different rates: the larger makes one planetary rotation each 16.2 h, while the smaller, more faint cloud lags behind, taking 16.9 h to circle the planet. The clouds are probably methane ice (NASA/JPL)

Sandia Laboratory's Z machine simulates conditions at the core of gas giants (photo: Randy Montoya; Sandia National Laboratories)

siblings has been a heated topic of discussion among atmospheric researchers. Before the Voyager encounters, some theorized that high altitude hazes were responsible, but even Earth-based telescopes seemed to be seeing deep into the atmospheres of the ice giants. On both planets, there was plenty of material to condense in the form of methane. At the temperatures in their neighborhood, methane in the upper atmospheres of Uranus and Neptune should condense into mist, leading to clouds. In fact, the few discrete clouds that have been observed on the ice giants appear to be condensed methane, but they are rare. Where are all the clouds?

"It's really weird that on both Uranus and Neptune, we don't see gigantic zones of clouds like we see on Jupiter and Saturn," says Kevin Baines. "It's one of the most perplexing things about these two planets. We've been trying to explain it away for a couple of decades."

In any atmosphere, a battle rages between upwelling air currents that keep particles afloat vs. gravity pulling particles down. For clouds to form, particles must stay airborne long enough to assemble into a group large enough to become visible. Larry Sromovsky, researcher at the University of Wisconsin-Madison's Space Science and Engineering Center, has been studying this battle as it relates to the atmospheres of the ice giants. "If there is not much vertical mixing, sedimentation will win even if there is not a rapid formation of big particles." Another factor, says Sromovsky, is the abundance

Beach ball-sized raindrops of methane accompany an atmospheric probe as it descends toward Neptune's lower cloud deck. Such raindrops may exist on both Uranus and Neptune (© Michael Carroll)

of what are called condensation nuclei, small particles that act as cloud seeds, attracting liquid to form droplets. If the nuclei are few and far between, larger droplets will condense, falling more rapidly and forming fewer clouds. But if the nuclei are more abundant, says Sromovsky, "smaller particles can form and persist longer." Both planets have condensing clouds, but not as many, on average, as models indicate they should have in a well-mixed atmosphere.

Kevin Baines advocates yet another possibility for the missing clouds. "The best theory suggests that when methane condenses, it condenses so rapidly, and there's so much of it, that over a few seconds to a minute you go from a little droplet that grows to the size of a beach ball, and the beach ball falls. You don't see any clouds, because it all rains out of the atmosphere too quickly."

Baines believes these super-raindrops may result from the fact that the ice giants are so rich in methane. Their atmospheres contain 20 times the amount on Jupiter and Saturn, and in the surrounding environments, the methane can condense, just as water does on Earth. But methane is a large molecule, and Baines sees this as significant. "This condensable is heavier than the gas it's in. The gas is hydrogen and helium, and the methane is 8 times heavier."

The Hubble Space Telescope came on line in 1993,[1] 7 years after the Voyager encounter. As Uranus has moved along its 84-year orbit, it has made its way toward equinox, the period when the north and south hemispheres receive the same amount of light. Equinox occurred in 2007, with sunlight illuminating the entire planet. Many scientists suggested that a polar hood similar to the one in the south would appear in the north as well. Others anticipated increased cloud activity within the green giant's deep atmosphere.

1. HST was launched in 1990, but because of a flawed mirror, observations didn't get started in earnest until a daring repair mission was carried out in 1993.

A dark spot on Uranus, imaged by the Hubble Space Telescope, showed Uranus's meteorology to be less tame than originally thought (NASA, ESA, L. Sromovsky and P. Fry of the University of Wisconsin, H. Hammel of the Space Science Institute, and K. Rages of the SETI Institute)

5,000 miles
8,000 kilometers

Uranus Dark Spot
Hubble Space Telescope • Advanced Camera for Surveys

NASA, ESA, and L. Sromovsky (University of Wisconsin)   STScI-PRC06-47

Day has finally dawned in the north, and observers are watching for differences between the northern and southern regions, says Larry Sromovsky. "There has been a bright band developing in the northern hemisphere, near 40–50° North, and a fading of the bright band in the southern hemisphere. There is no obvious sign yet of a polar hood developing."

Kevin Baines and others find the clear northern region puzzling. Baines suggests that polar hazes may be a seasonal effect. "It may be the case that once you have the pole lit up by the Sun and heated for many decades you build up this polar cloud. Maybe Uranus hasn't had enough time yet." Observers believe that if the hood is a seasonal feature, they will see changes within the next few years. For now, Uranus appears to be the most peaceful of all the gas giant worlds.

Still, Uranus has shown some dramatic changes since the Voyager encounter. Heidi Hammel has been studying both Uranus and Neptune using two powerful instruments: the Hubble Space Telescope and the Keck 10-Meter Telescope atop Hawaii's Mauna Kea. One aspect of her studies involves the heat flowing from the Uranian poles. Those temperatures have been very nearly identical, Hammel says. "The fact that they were comparable tells us that there's a lot of redistribution of energy within the upper layers of the atmosphere of the planet. The fact that the sunlit side is not warmer tells us there's some kind of a phase lag in the atmosphere. It takes a pretty long time for the atmosphere to fully respond to the extreme changes in sunlight that Uranus experiences because it's tipped over so far on its side." Given that Voyager data suggested this long phase lag, researchers expected to see very little change in the planet's clouds for years to come.

Instead, Hammel found features that resembled the diversity of Neptune's meteorology. "We saw lots of clouds popping up all over Uranus. We saw a lot of bright features appearing, poking up to high altitudes. The Hubble and Keck Telescope images showed a dark feature along the lines of Neptune's Great Dark Spot. So there were all sorts of transient rapid cloud activity, which doesn't seem to jive with a very long phase lag. We're all still puzzling over that. We're waiting for the modelers who love to delve into that kind of stuff."

The dark spot on Uranus reinforced the idea that the weather on the green giant is becoming more similar to that on Neptune. Although dark storms last for roughly 5 years on Neptune, the duration of those on Uranus is unknown because of limitations on earthly schedules. Hubble Space Telescope and the Keck 10-Meter are the only two facilities in the world with enough spatial resolution to detect these features, and time on those two facilities is limited, Hammel explains. "With Hubble, we might get 6 h a year. With the Keck we might get 10 or 12 h a year. So we try to piece together information with that limited amount of temporal coverage. That's just the reality of it. We're really pushing the edges of what can be done." Despite these limitations, Hammel could tell that the dark spot was a comparatively long-lived storm. She could see it or companion bright clouds associated with it for many months, possibly as long as a year. "It wasn't as big [as a similar dark spot on Neptune], but it did have companion clouds, so it seems similar structurally. Our spatial resolution is just horrible, although it's the best you can do in the world. All we can say is there was kind of a dark spot and there were kind of bright clouds around it, but we can't get any of the details like we got with Voyager. You're talking about a collection of 30 or 40 dark pixels surrounded by a few brighter pixels."

Voyager left scientists with a picture of clear skies and quiet meteorology on the green ice giant. Subsequent work by Hammel and others contradicts this. As Hammel points out, her observing runs have, for example, captured "clouds penetrating to high altitudes and subsiding on very short timescales. Some clouds are among the brightest ever seen in the Solar System."

Although its weather appeared, to Voyager, more peaceful compared to the other gas giants, the interior of Uranus is anything but. The planet's unique magnetic field gives clues as to what secrets hide beneath Uranian clouds and hazes. Researchers believe Uranus's inner core includes

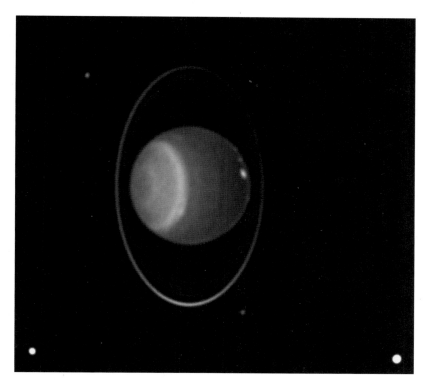

Hubble sees deep into Uranus's atmosphere in this infrared image. Bright areas are warmest. (NASA/JPL/STScI)

Methane wells up from the lower deck on Uranus in this artist's concept. Once an anvil forms, high altitude winds shear the cloud into a long structure called a tadpole (© Michael Carroll)

magnesium silicates and iron, a likely place to generate a magnetic field. But the bar-magnet-like field within Uranus is offset from its center so far that it may have nothing to do with its rocky core.

Two different scenarios could explain the odd energy fields surrounding the ice giant. The first, considered the most likely, is that highly pressurized water generates the magnetic field. Oceans on Earth generate similar fields in addition to those coming from our iron core. The other possibility is that the planet's core is undergoing a magnetic reversal. Magnetic stripes in Earth's expanding ocean floors document such changes here, and the same phenomenon may be in progress at Uranus.

Recreating the Heart of a Giant

For nanoseconds at a time, the heart of a gas giant beats right here on Earth. This mind-numbingly hot, superpressurized realm floats within a lightning-laced chamber at Sandia Laboratories in New Mexico. The chamber houses the "Z-machine," the most powerful electrical device in history. It is 108 ft in diameter, arranged in several concentric "doughnuts." Massive 90,000-volt capacitors stand in a 20-foot deep pool of oil, which rings an inner pool of deionized water. Devices submerged within this toroidal pool amplify the capacitors' energy, channeling it into a central vacuum chamber at the core of the structure. A 50 trillion watt electrical pulse plows into its center. In 1/10 of a millionth of a second, Sandia technicians generate 80 times the entire power output of the world. The tremendous electric current floods into a column of fine wires. The power vaporizes the wires, but for a moment the magnetic field surging around the delicate assembly is held in place. Then, the field collapses, driving atoms into each other. Pressures at the central core can reach 3.25 million atmospheres, with temperatures soaring above 50 million degrees Celsius.

Researchers believe these are precisely the conditions that may exist in the interiors of the gas giants – and in particular, Jupiter – where hydrogen is stripped of its electrons and becomes, in effect, a liquid metal. Scientists hope the Z-machine will provide a window on the structure of Jupiter's internal workings. "Superficially Jupiter has a very turbulent atmosphere," says Caltech's Andy Ingersoll. But its zones and belts are extremely stable. They have remained essentially unchanged for a century. "Do the bands have deep roots? It's possible. The belts may be arranged as nested cylinders that are concentric to the axis of rotation. This would set up tremendous centrifugal forces. When you start moving that sort of mass around, it affects the gravity field of the planet."

Engineers who run Sandia's Z machine work with scientists such as Ingersoll who are attempting to model the interiors of the gas giants with super-computers. Their complex models of planetary structure provide the basis for Sandia's experiments. Scientific theories make predictions. The engineers then design an experiment, and the scientists compare the Z-machine's results to their predictive models. The models may be simple compared to reality; researchers have so little information that they cannot know the details. But armed with tools such as the Z-machine, they are confident that our knowledge of Jovian cores will soon be revolutionized.

NEPTUNE

In 1989, while scientists continued to ponder the results of Voyager's Uranus encounter, the spacecraft reached Neptune. It had been a 12-year journey fraught with near disaster on several occasions (see Chap. 3).

Neptune's remarkable blue face stands in stark contrast to the soft green of Uranus. Its clouds are intrinsically blue, says Heidi Hammel. "We've done comparative spectroscopy of Uranus and Neptune, and the slope of Neptune is quantitatively more tilted toward the blue on Neptune. It's a sort of straight line, which is telling us that there is some kind of coloring agent in the atmosphere that gives it the more bluish color compared to Uranus. All of the clouds are blue; it's not just the dark spots." The reason both ice giants are blue-green is because they are missing red light. Three percent of the atmosphere weighs in as methane, and methane molecules are incredibly efficient at absorbing certain wavelengths of light in the red, Hammel says. "When you look at the spectrum, instead of being a flat line, there are these big chunks taken out of the spectrum at the red end; it's missing red light. When you take away red light, what have you got? A lot of blue. But you'd think that they'd both be the same blue, because methane is absorbing the same reds. But Neptune has this extra coloring agent." What that agent is remains a mystery.

As the blue giant gradually grew from a bright point of light to a disk, Voyager's imaging system resolved amorphous blobs into belts and spotty clouds. Even in the earliest of images, it became obvious that Neptune harbored far more active weather than Uranus. Voyager monitored Neptune's unique ring arcs and, in the final days of encounter, mapped its moons. One of Voyager's most significant atmospheric discoveries was that – like Jupiter

The disk of Neptune broods in the dim twilight of the outer Solar System in this Voyager 2 image centered on the GDS. Scooter and D-2 are visible near the left limb of the planet (NASA/JPL)

and Saturn – Neptune puts out more heat than it receives. Kevin Baines explains, "Jupiter is warmed up significantly more than it should be. Saturn is warmed up even more. Neptune gets only half the sunlight that Uranus does, but it warms itself up to be the same temperature as Uranus."

The internal heat makes for the dynamic activity in Neptune's weather. White clouds of methane crystals skitter across subtle belts and zones, sometimes spiraling into cyclonic features. Unlike the clouds on Jupiter and Saturn, Neptune's were difficult to track, says Larry Sromovsky. "The bright discrete clouds on Neptune evolve fairly rapidly, so that clouds seen on one night cannot easily be identified on the next rotation." This makes it difficult to analyze the dynamics of Neptune's atmosphere. Kevin Baines underlines the challenge facing outer

planets researchers. "You go to Jupiter and see the Great Red Spot, and it's there forever. You see zones and belts that, on the first order, are always there. You see changes on yearly scales and a few abrupt shifts. But on Neptune it's all the time. There are no real zones; you get places that look like they are half-filled zones, and you've got those wispy clouds that are very ephemeral – they come and they go. It's very hard to measure wind speeds on Neptune because you can't tell if the cloud you're following is the same cloud that you saw an hour before."

Like the other gas giants, Neptune's atmosphere is dominated by hydrogen, the most abundant element in the universe. The second most abundant gas is helium. Like hydrogen, helium is essentially clear in visible light. It is the third gas, methane, that gives Neptune its distinctive blue. Deeper down, below the blue cloud deck, the atmosphere is likely infused with water and ammonia. In Neptune's clear upper atmosphere, at a pressure of about 100 mbars (1/10th that of Earth's surface pressure), temperatures hover at about −360°F. Descending from this level, temperatures rise with depth. At a pressure comparable to Earth at sea level, temperatures reach −334°F. Beneath this level, the air is warm enough for methane to exist as a liquid or vapor. The methane ice crystal cloud deck has its base at a pressure of about one bar, where methane can cool and condense into clouds. The clouds billow up above this altitude, bringing methane into the stratosphere. But Neptune's stratosphere has more methane than it should be able to hold at its temperature, explains JPL's Kevin Baines. "Neptune's stratosphere has an excess of methane vapor that must have been transported vertically upward from the depths in powerful methane-cloud storms. These storms break through the tropopause and sublimate in the high stratosphere, depositing their methane vapor. Such storms must traverse something like 50–100 km of altitude."

For the past three decades, scientists theorized that the large amount of stratospheric methane vapor was due to this "convective overshoot": extremely powerful storms billowing up into the lower and middle stratosphere, carrying methane ice particles that then cool, condensing into ice particles. But recently, in light of new Earth-based observations, researchers have postulated that the pole of

Voyager 2 captured this view of methane clouds just 2 h before its closest approach to Neptune, at a range of 98,000 miles. The clouds roughly parallel the equator. The streaks are between 20 and 125 miles across, and float above the "blue-tinted cloud" layer at an altitude of some 30 miles (NASA/JPL)

Neptune may have an unusual thermal structure, which would allow methane to leak up into the mid-stratosphere levels. This avoids the need for such strong global methane convection, which some scientists doubt. But Kevin Baines cautions, "The jury is not completely in yet on this idea." Uranus does not show the excess of methane in its stratosphere that Neptune does.

Despite the transient nature of its clouds, Neptune does have a few long-lived features. One of the first features Voyager resolved in far encounter shots was a large, blue storm reminiscent of Jupiter's Great Red Spot. The storm was approximately the same size relative to the planet, a deep pool of sapphire floating in a cobalt sea of clouds. Scientists quickly named it the "Great Dark Spot" (GDS-89), a reference to Jupiter's Great Red Spot.

As Voyager resolved more detail, the spot took on its own unique character. The Great Dark Spot spanned the distance of Earth's diameter. This oval-shaped storm appeared to rotate or oscillate over an 8-day period, and researchers came up with a model to explain the storm's behavior. Computer simulations indicate that

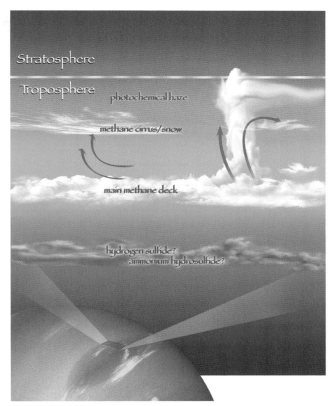

The atmospheric structures of the ice giants differ from that of Jupiter and Saturn (© M. Carroll)

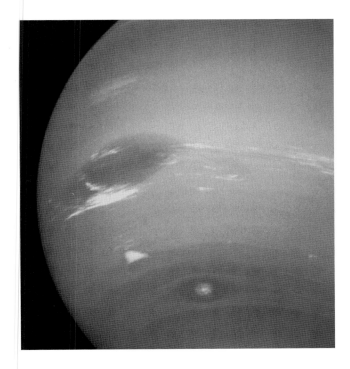

This Voyager 2 photograph of Neptune captures a moment when three distinctive Neptune features lined up on the same hemisphere. The Great Dark Spot (top left), the white cloud Scooter near center, and the smaller dark storm D-2 all moved at different eastward rates, rarely appearing close to each other (NASA/JPL)

Three images of the Great Dark Spot, taken at 18-h intervals, reveal changes in its bright, cirrus-methane clouds. These changes are dramatic, considering their scale: The GDS is about as far across as Earth (NASA/JPL)

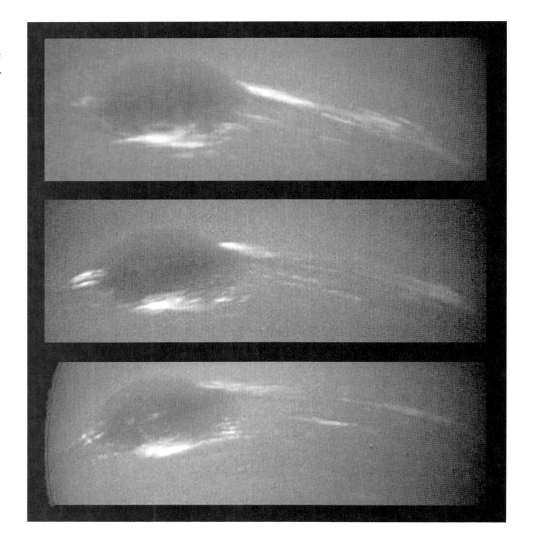

GDS-89 was a vortex turning anticyclonically within the streams of Neptune's surrounding atmosphere. The storm acted as an obstruction in the wind's flow. As air was deflected upward, methane clouds condensed. This type of cloud is known as an orographic cloud and is often seen on the leeward side of mountains on Earth. In fact, bright methane clouds did cling to the edge of the storm. But the direction and the nuances of GDS-89's movements proved difficult to judge, says Larry Sromovsky. "The detailed motion of clouds within the vortex, to verify that it actually spins with the direction and amplitude of our model, could not be verified by observations. Trackable cloud features were too sparse and too variable to allow tracing the atmospheric flow."

Sromovsky's computer simulations provided insights into the storm's altitude, which had been debated in the months following Voyager's encounter. The model shows the top of the storm to be within the tropopause, based on comparing the model with the way the GDS migrated across the face of Neptune. According to the final report published in *Icarus*, "When a GDS is

started with its top in the stratosphere it drifts much too rapidly
toward the equator and quickly
disperses. On the other hand, if its
top is well below the tropopause
there is a tendency for the companion clouds to be too large. Hence
the top of a GDS is probably at the
tropopause."

By the time the Hubble Space
Telescope came on line in 1993, the
Great Dark Spot of 1989 – a massive, planet-sized cyclone – had
vanished without a trace. But in
1994, Heidi Hammel spotted a
similar storm in Hubble Space
Telescope images of the northern
hemisphere. Now known as GDS-
94, this storm was slightly smaller.
It disappeared within several years
and was followed by others. After

The most detailed view of a small dark spot called D-2. Bands wrapping around the feature are evidence of strong winds, while structures within the bright spot may indicate both upwelling clouds and rotation of the dark spot around its center (NASA/JPL)

the disappearance of the 1994 storm, Hammel and her team discovered two
more. "They seem to last about 5 years or so. The 1989 one probably formed
in 1988 or so, based on its companion structures – the bright clouds that
accompany it. In 1994 we got our first Hubble images, and we saw that big
dark spot in the north. After about 5 years, that one was pretty much gone
and there was another one a little closer to the equator in the north, but then

Neptune Clouds
PRC95-21A · ST ScI OPO · April 19, 1995 · H. Hammel (MIT), NASA

HST · WFPC2

The Hubble Space Telescope took these three images of Neptune in October of 1994. Neptune was 2.8 billion miles away. High-altitude methane ice clouds appear bright in these near-infrared images (NASA/JPL/STScI)

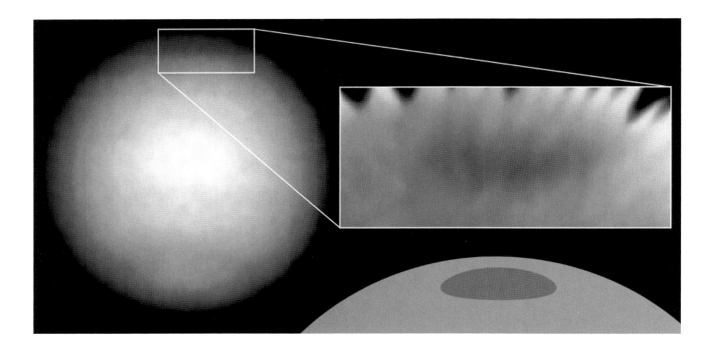

*Hubble image of Neptune's
dark spot GDS-94 (Space
Telescope Science Institute;
Heidi Hammel, MIT; and
NASA)*

that one disappeared after about 5 years and we haven't seen one since.
We've had sort of a dry spell when it comes to dark spots on Neptune."

Voyager observed other bright clouds drifting on Neptune's alien winds.
The white clouds on both Uranus and Neptune usually consist of thin parallel filaments. At Neptune, some stretched in wisps for hundreds of miles,
while others skittered over the deep blue lower cloud deck. One such cloud
was nicknamed Scooter. Unlike most of the bright methane clouds, it was
embedded deep in the atmosphere. Like the GDS and other cloud features,
the cloud moved eastward. It appeared to move quickly compared to the
GDS, inspiring its name. "The Scooter was named when we didn't really
know what the interior rotation rate of Neptune was," says Sromovsky. "It
actually moved relatively slowly, about 1° of longitude per hour compared to
the GDS, which moved over 2.5° per hour. The Scooter might be a methane
cloud, just not as high as others that show up in methane band images."
Whether it is a convective feature, such as a thunderstorm, or a wave feature
similar to a lenticular cloud on Earth is not known. But Sromovsky has his
money on an updraft from the rich methane reserves below. "I would bet
that upwelling of some sort is needed to produce a region of enhanced condensation, either waves or vertical convective mixing."

The dramatic weather systems on Neptune confounded some predictions. Many meteorologists believed that atmospheres of planets far from
the Sun would be comparatively calm. With less heat to drive air movements,
they assumed that Saturn's winds would be more subdued than Jupiter's,
and that Uranus and Neptune would continue the trend toward quieter
weather. But distance does not bring calm, says Caltech's Andy Ingersoll.
"The idea we get is that the winds don't decrease as you move out in the

Solar System. That was a significant finding. It was borne out as Voyager moved on to Uranus and Neptune. That's quite dramatic for Neptune because it gets only 5% of the amount of energy that Jupiter does on a per area basis. All the giant planets have stronger wind fields than Earth. Why should it be that the winds are stronger in the outer Solar System? Even among the giant planets, why should they be stronger at Neptune than at Jupiter?"

Ingersoll thinks he knows the answer. As the amount of sunlight heating a planetary atmosphere is reduced, several things happen at once. The power that drives the large-scale jets and storms diminishes, but so does the small-scale turbulence. "The turbulence ultimately limits the speed of the jets as well as keeping the jets going. So the dissipation wins out as you move into the outer Solar System. As you move out to Neptune, you produce an atmosphere that has so little small-scale turbulence, so little dissipation, that the winds are free to build up and get stronger. What makes atmospheric dynamics a hard field is that you have all these different phenomena at different scales, different sizes, all interacting with each other. That's part of the fun."

Neptune had yet another puzzle for the scientific community. As the planet moves in its orbit and its distance from the Sun increases, it seems to be warming. In fact, Neptune's atmosphere has been steadily warming over the past 20 years. Hammel is intrigued. "What I make of it is that I don't think we understand our atmospheres in the Solar System as well as people think we do. It does raise questions in my mind about how well we understand temperature warming in atmospheres and where it might be coming from in the Solar System. You have to be a little careful about rash assumptions that we understand why atmospheres warm, and I'm not limiting it to the outer Solar System. There may be mechanisms that we are not fully accounting for as we model planetary atmospheres. That includes Earth."

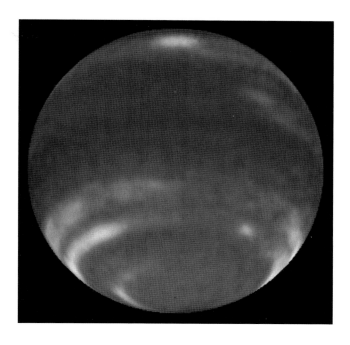

From 1996 to 2002, Hubble images showed a dramatic increase in reflective white cloud activity on Neptune. This 2002 image, looking up toward the south pole, shows bands and distinct clouds (NASA/STScI)

Getting High: How Clouds Condense in Stratospheres

The bright clouds that drift in the skies of Uranus and Neptune are not the same kinds of clouds that swirl across Jupiter and Saturn. On Jupiter and Saturn, light clouds are typically ammonia. On Uranus and Neptune, bright clouds appear to be mostly methane. Methane is very difficult to identify spectroscopically, so scientists must use other clues to determine what those clouds are. Methane is the strongest suspect from a thermodynamic perspective. Researchers know how much methane gas is present at a given altitude, and they know what the temperature at that altitude should be. Methane does condense out in these conditions, and clouds appear right where they should. The pressure at the level of the clouds also confirms that the clouds are likely to be methane.

Clouds form when vapor condenses and becomes visible. The supply of condensable vapors in planetary atmospheres comes from below. For water clouds on

Earth, the oceans provide the primary source of condensable material. For the gas giants, the well-mixed part of the atmosphere below the clouds contains cloud-making vapor.

Normally, condensable vapors (like water on Earth) remain in the lower part of the atmosphere, within the warm troposphere. At this level, the temperature is too warm for condensation to occur. At higher altitudes, temperatures drop. Vapors condense and form clouds as they climb into the cooler part of the upper troposphere. Rising hot air, convection, is by far the most powerful process for transporting material upward. But, in the uppermost troposphere, just below the tropopause, temperatures level out and convection shuts down. Above the tropopause, temperatures actually increase with altitude – which puts an even bigger damper on convection. All of these factors conspire to keep most of a planet's condensable material in the troposphere, below the stratosphere.

Some exceptions to this are cirrus clouds on Earth and in Neptune's stratosphere, where there is an abundance of methane. On Earth, especially powerful thunderstorms can punch through the damping part of the troposphere, thrusting water vapor into the stratosphere. Neptune seems to be an extreme example of this, with high amounts of methane in its stratosphere.

Cryovolcanic geysers break through a surface layer of nitrogen ice on Neptune's largest moon, Triton. The plumes rise through calm air until they reach a jet stream 8 km above the ground, where they are sheared off toward the north (© Michael Carroll)

TRITON

Flight planners considered Neptune's large moon Triton to be one of the top priorities of Voyager's 12-year grand tour. Measuring 2,706 km, it's the seventh-largest moon in the Solar System. Early studies hinted at traces of a tenuous atmosphere. Perhaps more important was the fact that Triton may be a captured wanderer from the outer fringes of our Solar System. Its orbit is retrograde; it orbits Neptune in the opposite direction of Neptune's spin. If Triton had been formed within the Neptune system, it should circle the parent planet in the same direction, and roughly in Neptune's equatorial plane. But Triton's orbit carries it in the opposite direction at a dramatic tilt.

The possibility that Triton came from further out captivates scientists. Beginning at about the orbit of Neptune lies a torus-shaped cloud of comets and ice dwarf planets, some larger than Pluto. These Kuiper Belt Objects (KBOs) have been the subject of intense study in recent years, and Triton may provide a glimpse of what these icy worlds are like. One such world, Pluto, is nearly identical in size to Triton.

Voyager images of Neptune's largest moon confirmed that the little world is one of the strangest places in the Solar System. As we have seen with the terrestrial planets and Titan, the surface below informs us of the atmosphere

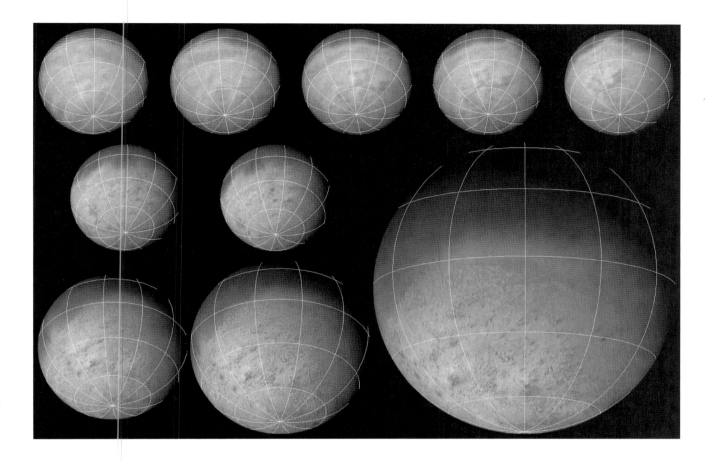

above. If Triton's landscape is any indicator, bizarre things are afoot in the air. As Voyager 2's high resolution images appeared for the first time on screens at NASA's Jet Propulsion Laboratory, a hush settled across the press room. Gasps could be heard amidst mumbled words such as "weird" and "creepy." Triton's surface spread before the scientists and observers as a tortured battlefield of pockmarks and fissures. But what battle was raging there? Few craters were evident, implying that either the ground was being resurfaced from internal forces or obliterated by Triton's weather. Flat plains seemed to overflow hummocky terrain. Toward the equator, overlapping ridges and circular depressions earned the nickname "cantaloupe terrain." Many features continue to perplex geologists, 20 years later, says Dave Stevenson. "I think we still don't really understand Triton, but certainly the morphology, this cantaloupe terrain, seems to require some sort of internal process. That's not what you get with craters. Craters come in all sizes, whereas the cantaloupe terrain has a particular size associated with it which would suggest some sort of internal process."

Toward the south, dark material pooled in basins, reminding analysts of lakes. A series of dark streaks led away from the south, spreading out as they faded toward the equator. The streaks seemed to emanate from Triton's nitrogen ice cap, a ragged pink mottling that seemed to be sublimating at its frayed edges. What was the surface trying to tell us?

This magnificent mosaic of Voyager's best images shows Triton's south polar cap on the left side of this image. Pink nitrogen ice is in the process of sublimating into the thin atmosphere (NASA/JPL)

The weather seemed an unlikely culprit for the moon's wound-like features, as Triton's atmosphere turned out to be far thinner than most expected. While Mars's atmosphere is about 8/1,000 that of Earth's, Triton's was a thousand times less than that of Mars (15 µbars) during Voyager's 1989 encounter.

The other possibility – that internal forces had sculpted the surreal landscape – was reinforced by the edges of the smooth plains, which are often lobate and resemble flows. In several locations, the flows appear to come from a central pit or collapsed region, further evidence of cryovolcanism (super-cold volcanic activity). But what of the mysterious streaks? The dark trails seemed to be wind-related. Could Triton's weak meteorology be responsible?

Triton's air consists mostly of nitrogen. Trace amounts of methane are photodissociated into hydrocarbon hazes. Because of Triton's low gravity, the outer fringes of Triton's atmosphere waft at an altitude of 800 km above the frigid surface. Like terrestrial worlds, the base of the air blanket has a troposphere where the weather phenomena tend to stay. The troposphere tops out at about 8 km up. Triton lacks any kind of stratosphere, but instead, its troposphere grades directly into a thermosphere, ionosphere, and exosphere.

If the streaks across the polar ices were weather-related, Voyager scientists reasoned, that weather must occur in the troposphere. It was clear that Triton's feeble atmosphere was somehow tied to the polar ices, which were also composed of nitrogen. Scientists scoured the few dozen images of Triton

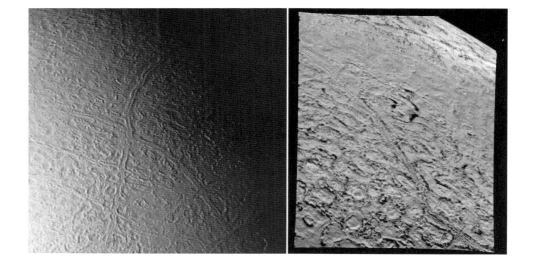

Two images of the peculiar "cantaloupe terrain." The image on the right is the highest resolution view ever seen of the icy moon (NASA/JPL)

that Voyager was able to snap during its brief flyby. The portrait that scientists assembled of Triton's weather is truly an alien one. Triton's day lasts nearly six Earth days, and because of Triton's retrograde motion, the Sun sets in the east. Triton's surface temperatures hover at about −235°C, making it the coldest surface visited in the Solar System. About 800 km up, at the top of the exosphere, the 4.5 billion-km-distant Sun heats the air to a balmy −180°C. Near the surface in the southern hemisphere, winds blow to the northeast (as seen by the dark streaks). Winds spiral around the pole in a great anticyclone, unwinding toward the equator and reaching velocities of 5 m/s. They are energized by the difference in temperatures between the polar ices and the warmer exposed ground near the equator. (This "ground" is water-ice and is warmer and darker than the nitrogen ice.)

Between 1 and 5 km altitude, thin nitrogen clouds occasionally condense. Over the course of Triton's 165-year orbit, the clouds may become dense enough for snows to fall. High altitude westward winds stream across the tiny world at the top of the tropopause, around 8 km above the surface. This jet stream was dramatically underscored in several Voyager images that

Dark areas called "lacus," Latin for lakes, may be holes melted through the nitrogen ice cap. If so, these features will be dynamic and changeable (NASA/JPL)

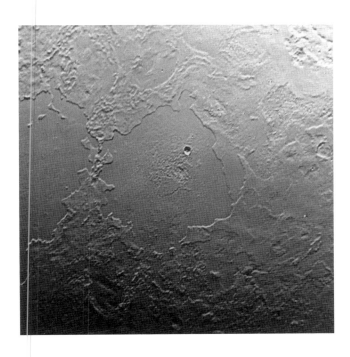

This basin resembles a volcanic caldera, with its steep walls and collapsed central pit. If erupting material from Triton's interior led to this formation, cryovolcanism must contribute to Triton's atmosphere (NASA/JPL)

captured at least four active geysers. The dark nitrogen plumes, related to the polar streaks, rose vertically into Triton's thin air, then made an abrupt dogleg at the 8 km high wind shear.

Winds, clouds, hazes, even air pressure – all the elements of Triton's weather – are driven by its polar cap. At the time of the Voyager encounter, the southern hemisphere was in the last quarter of its 40-year-long spring, and the polar cap was presumably shrinking. Nitrogen from the frozen cap was sublimating into the atmosphere, adding pressure. In some ways, Triton's nitrogen ices and atmosphere are reminiscent of Mars's seasonal carbon dioxide cycles between its atmosphere and polar caps. In both cases, frost on the ground absorbs sunlight in the spring and gives off vapor. Andy Ingersoll likens this process to sweating. "The heat from the Sun goes into evaporating the vapor, which is exactly what happens to us on a hot summer day. Our temperature doesn't rise, because we give off vapor that takes away the heat. On Mars, it's CO_2, and there's an annual deposit of CO_2 that's put down in the winter time. When the Sun hits it, it evaporates and it goes into the atmosphere. The same is true on Triton, except that the ice is nitrogen."

There is an important difference between the two planets: on Triton, as winter arrives at the opposite pole, the nitrogen on Triton migrates there. In a sense, Triton's atmosphere is simply a gaseous step in the migration of the ice caps from one pole to the other. Unlike Mars, Triton does not have enough nitrogen to support an atmosphere all year, says Ingersoll. "The nitrogen atmosphere of Triton, according to Voyager, is roughly a thousand times less than on Mars, and the annual cycling of nitrogen is in the same ballpark as the annual cycling of CO_2 on Mars. So instead of that being 30% of the atmosphere, on Triton it's the whole atmosphere. The CO_2 on Mars is capable of supporting a 7-mbar atmosphere, which is really fairly substantial. At any one time, there is more carbon dioxide in the atmosphere than there is in this annual cycle of carbon dioxide. The atmospheric pressure goes up and down on Mars by about 30%. What's happening on Triton is that the whole atmosphere collapses twice a year, when it's winter time on one pole or the other. It blossoms out in the springtime and then collapses again. Instead of having a 30% annual cycle, it's 100%." Triton only has "weather" during the spring and fall, because it has an active atmosphere only during those seasons. Dave Stevenson adds, "Nitrogen is the dominant atmospheric constituent. Nitrogen is mobile, and in that sense it's playing the same role that methane plays on Titan. Triton is much colder, and that's why you shift the emphasis for the atmosphere and material on the surface to be nitrogen."

The geysers of Triton may be related to the sublimating ice in the polar caps. They appear to break through the nitrogen ice, with surface trails extending from the source for over 100 km. The plumes themselves rise up to 10 km in narrowly focused columns, much like steam rising from a smokestack in calm air. Estimates of the diameter of the columns range from several tens of meters to 2 km.

The dark color of the eruptions may provide a clue to their composition. When nitrogen is irradiated by sunlight, it turns a dark brown, similar to the plumes and drifts of material draped across Triton. It seems that the geysers are gaseous forms of the ice from which they come. Several ideas have been put forth for the genesis of these otherworldly geysers. The first suggests a "solid-state greenhouse effect." In this scenario, sunlight penetrates through the nitrogen ice, reaching a translucent layer of darker ice below. Heat begins to build up inside the ice, and networks of subsurface pockets of nitrogen gas form from the sublimating ice. Pressure builds up to the point where the ice fractures, and the nitrogen gas spews into Triton's thin air.

A second model describes a super greenhouse effect. Sunlight shines deep into the nitrogen ice (which is very clear compared to water-ice). At the base of the ice lies a dark layer of material that absorbs the heat, forming pockets of nitrogen as in the first model. Researchers estimate that several hundred kilograms of nitrogen gas erupt per second. Eruptions may last several years. If so, up to a tenth of a cubic kilometer of ice sublimates. A third hypothesis, put forth by Andy Ingersoll, is that sunlight heats patches of ground that are dark and free of frost. These areas would heat up more quickly than the surrounding nitrogen ice, setting vortices into motion. These Tritonian dust devils could carry particles up into the high troposphere.

Active plumes drift aloft in a different direction from the trails left on the ground. The reason has to do with the outflow along the surface from the subliming south polar cap, which sets up a different wind direction from the high-altitude wind shear caused by the warm equator and

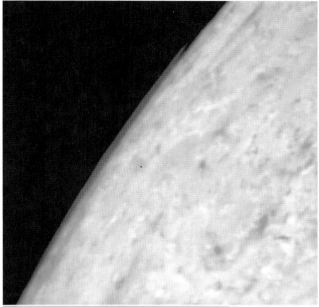

Clouds over the limb of Triton's south polar cap. The cloud extends 100 km along the horizon. The images have been stretched to bring out detail; the horizon is indicated by the dotted line in the lower frame (NASA/JPL)

cold south pole. Ingersoll compares the situation to that on Earth. "Triton spins in a right-hand sense about the south pole, which makes it like Earth's northern hemisphere. The winds aloft are like Earth's jet streams, and the winds near the surface are like the outflow from a northern hemisphere high-pressure system. The difference is that the high pressure is centered on the pole due to sublimation, and Earth doesn't have an analog to that."

Stevenson believes there may be another culprit behind Triton's cryovolcanism: methane. "The idea that people have in mind is that underneath this

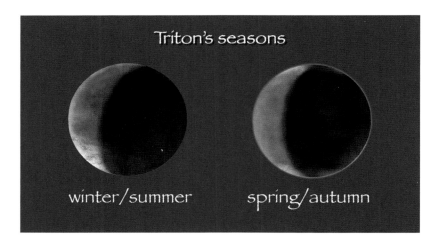

The seasons on Triton are tied directly to its meteorology. During winter, the atmosphere collapses completely, blooming again in spring and fall as it migrates from one pole to the other

thin veneer of nitrogen ice and possibly liquid, there is a layer of solid methane. Were you to get sunlight down to that layer, you may heat it to the point where it would start to sublime and you would have a geyser. Nobody knows for sure how thick the nitrogen on the surface of Triton is, but the requirement from the spectrum is for a fairly modest amount." But methane might be a candidate for those mystery geysers, Stevenson suggests. "The surface temperatures on Triton are so low that methane is very happy as a crystal. But if you're bringing in energy to subsurface, then you can heat the temperature in those subsurface regions up to the point where methane ice wants to make a significant amount of vapor. That doesn't require a really large temperature increase."

With the nature of Triton's geysers still up in the air, researchers were anxious to see if Triton's atmosphere would go through changes in the years after Voyager's flyby. In the 1990s, observatories trained their telescopes on stars being occulted by Triton. As the starlight dimmed, scientists could tell that the moon had a denser atmosphere than at the time of the Voyager encounter. Temperatures also seemed to be on the increase, says Heidi Hammel. "What's been going on is that the temperature has been increasing and the pressure has been dropping. It's possible that this is simply regular seasonal variability as the sunlight changes on Triton. It heats and warms the ices, and the volatile stuff in the ices just sort of migrates around the surface of the moon. We saw these cryovolcanoes, and we know that it's a dynamic, changing atmosphere." But that dynamic atmosphere begs the question: if the entire atmosphere collapses in winter, and if Triton is on its way to that season, why do we see these trends? Hammel and other researchers are mystified. "That's a really good question that we puzzle over and don't have a good explanation for. It's the same with Pluto. It's getting farther from the Sun, and you'd think its atmosphere would be getting colder. Wrong. Stellar occultations show it getting warmer."

Dave Stevenson wonders if the bizarre terrain and alien meteorology of Triton could be caused by internal heat far below the moon's crust, perhaps generated by radioactive decay. "Triton is a small body, but it's got a lot of

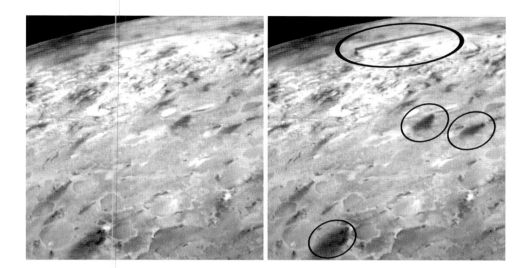

Voyager captured several eruptions in action. The image at right delineates one geyser near the horizon rising 8 km before hitting a wind shear that carries it to the right. Other plume tracks across the surface can be seen below it, with possible sources marked (NASA/JPL; overlay by author)

rock. From that rock comes radioactive heat. The model that one thing makes sense of for Triton is a rocky core with a layer of ice on top, and the ice on top is presumably water-ice, but there could perhaps be ammonia. There might be methane clathrate. In all those things, the heat source is radioactivity. If you want to do something interesting with volatile ices you don't need a lot of energy. Even low levels of radiogenic heat at these temperatures can have a profound effect on volatile-rich surfaces. It is, in general, much easier to do something with radioactivity in the outer part of the Solar System."

To Ingersoll, Triton's varied possibilities and its cacophony of interacting phenomena embody the beauty and challenge of science. Each planet, each new discovery, provides deeper insight into how the universe works. Looking back on the Voyager project with its many discoveries, Ingersoll says, "It's humbling to realize how limited our imaginations are, to see how varied and unpredictable the actual objects are. I guess any good meteorologist would have guessed some of these things, but we aren't all that good." If history is an indicator, future meteorologists will be dining on plenty more humble pie, served at the table of the outer planets.

Part III

Future Explorers

At the end of a 2-h flight, its fuel spent, an ARES robotic Mars plane descends toward its final resting place in the ancient canyons of Terra Sabae (© Michael Carroll)

Chapter 10

Field Tests, Balloons, Aircraft, and Upcoming Missions

In 1964, two engineers from Martin Marietta ran in circles through a Colorado wheat field with a handmade balsa wood and paper glider. The model was a perfectly scaled mock-up of a reentry vehicle they had been working on for several months. Their stubby "lifting-body" was a precursor to the modern space shuttles, and they wanted to see how it behaved before they sent a more expensive version into wind tunnel tests at NASA's Ames Research Center.

Field tests have come a long way since then. Today's engineers still tinker, but their wheat fields are often replaced by remote planetary "analogs," places that bear some geologic resemblance to other worlds. These include volcanic fields in Hawaii and Kamchatka, glaciers in Iceland and the Rockies, and remote arctic deserts in Antarctica and Canada.

SHRINKING TECHNOLOGY

Technicians recently completed a battery of field tests in the Canadian arctic using a miniature Unmanned Aerial Vehicle (UAV). Their staging area was the Flashline Mars Arctic Research Station, a facility run by the Mars Society. Though they labored in the Canadian wilderness, their eyes were on the Red Planet, says Joe Palaia, vice president of the private space development company 4Frontiers Corporation. "It's a good Mars analog for a number of reasons. Number one: it really looks like Mars. It's got polar topography, lots of canyons and hills. The habitat where I stayed in is actually located on the rim of the Haughton Impact crater. This particular impact crater is particularly well preserved. Elsewhere on the planet a lot of these craters have been eroded away, either due to action by water or wind or glaciers, and that's not the case for this particular crater. The other reason that it's a good place to go in terms of a Mars analog is because of its remoteness. In order to get there you basically have to fly as far north as you can on a commercial airliner. You can get as far north as Resolute Bay, which is a little hamlet up on Cornwallis Island. I think they only have about a hundred people that live there on a permanent basis. And then you have to rent a Twin Otter, one of these bush planes, and that takes you over about a half hour to a dirt strip on Devon Island. So the mindset that you're put in when you go to this very remote location is a similar mindset to what you might experience on Mars, where it's obviously quite remote. You can't

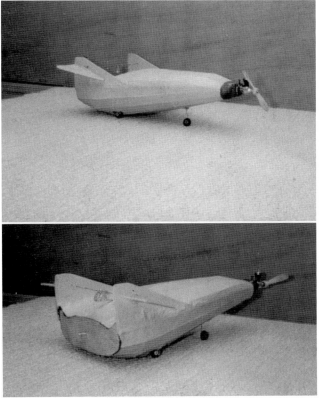

Balsa wood and paper model lifting body, ca. 1961 (from the collection of Patrick Carroll)

M. Carroll, *Drifting on Alien Winds: Exploring the Skies and Weather of Other Worlds,*
DOI 10.1007/978-1-4419-6917-0_10, © Springer Science+Business Media, LLC 2011

run out to the corner hardware store and get what you need if something breaks. You have to make do with what you have. From that perspective it's a good Mars analog, a lot better than, say, going right outside the door at Johnson Space Center to some topography there where if something breaks we can just run inside and get another bolt or run down to the store."

Within this environment, Palaia and others tested the latest in miniature UAV technology, called the Maverick. One of the most exciting areas of study for future atmospheric explorers involves remote sensing from deployable aircraft. Gainesville, Florida, company Prioria Robotics produces one christened the Maverick. It's an ultra light, carbon fiber structure that can fold up into a 6-in. tube. Palaia calls Maverick "a spectacular piece of technology." Unlike other deployable planet-exploring aircraft, Maverick needs no complex hinges or unfolding spring mechanisms. The patented wings and rudder unfurl as the plane is pulled from its carrying tube, Palaia says. "It looks very light and very strong and resilient to impact. With all the other UAV's you have to attach the wings to it. The wings are separate. In this case, it's a proprietary design of theirs where the wings actually bend and curl around the fuselage. And so you can store it with wings and all in this very compact tube until you pull it out of the tube and *pop*! The wings deploy, and you're ready to go."

For the Devon Island tests, Maverick had a Google map of the local terrain uploaded into its navigation system. A built-in GPS enabled the drone to relay its position back to controllers on the ground, who commanded it through Palaia's laptop computer. "You basically can program and modify its flight while it's flying. I can be dragging things around on my computer and then I can upload those commands to the UAV and then it can go and execute those commands."

Maverick would need very few physical changes to fly in environments such as Titan or the gas giants. Thermal and radiation considerations would be paramount in the frigid outer Solar System. On Venus, at atmospheric levels similar to those flown by the VEGA balloons, the craft would be fully operational as it is, as long as it was covered by anti-erosion materials to protect it from the harsh acid hazes of Venus. For Mars, the wingspan would need to be extended. While the Devon Island scenario included "astronaut" handlers to fly the craft, Palaia is confident that micro-UAVs such as Maverick have applications for robotic planetary exploration. "You could certainly fully automate it. You would have to have much more sophisticated software because the system is fairly dumb from the perspective of really understanding what it's looking at and where it is. It has some basic information about the Google maps and what's the terrain around it and at what elevation it thinks it is. But that elevation is above sea level, not necessarily the elevation above terrain. It will try to fly itself into a mountain if you put that on its flight plan."

Some advanced versions of Maverick are equipped with a forward-looking camera that incorporates hazard avoidance software. If the craft loses communications lock with the controller, it drops into a "time-out mode." After a certain interval, if no communications take place, Maverick retraces its steps,

Palaia says. "It will automatically turn around, head back to its landing zone, and if it still hasn't received communication it will go ahead and land itself. There are other safety features: if it detects some problem in maneuverability, if there's a problem with one of the flaps, or if the battery starts getting below a preset threshold, then the thing will turn itself around and come back home."

With miniaturization and advanced materials coming down the assembly line, Palaia and others feel that small-scale UAVs such as Maverick can either complement or replace larger robotic aircraft. Future planetary exploration may include swarms of micro-UAVs banking and gliding on alien winds.

Joe Palaia (right) consults his laptop control center as Kristine Ferrone (flight controller for the International Space Station) holds the miniature Maverick UAV. The flight was successful (photo © Stacy Cusack)

BIGGER BIRDS

The military has been operating UAVs, including the famous MQ-1 Predator, since 1995. These remote-controlled drones are versatile in the field and were originally used for military surveillance. They have since been modified to carry weaponry. Predator UAVs have the capability of flying 400 miles to a target, observing that target from the air for 14 h, and then returning to the base. The next generation of UAV is called the MQ-9 Reaper, which can cruise at 220 miles per hour for 16 h as high as 50,000 ft. Reapers can carry 1.5 tons of weapons. But what if a similar vehicle could carry that much scientific equipment through the alien skies of another world?

San Diego's General Atomics is modifying its Predator UAV for use by NASA to study Earth's atmosphere. The upgraded vehicle, called Altair, carries 700 lb of instruments at altitudes of 50,000 ft for up to 32 h. Engineers extended each wing by 11 ft, giving Altair an overall wingspan of 86 ft. A 700-hp rear-mounted turboprop engine propels the UAV, driving a three-blade propeller. Altair is capable of over-the-horizon control, collision-avoidance, and other technologies required to enable it to avoid other aircraft.

ARES: HOW TO MAKE A MARSPLANE

The automation of Altair can be applied to advanced systems that could explore distant worlds. One such project, in advanced stages of study and construction, is a UAV designed to fly above Mars. The ARES, for *Aerial*

*R*egional scale *E*nvironmental *S*urveyor, is under development at NASA's Langley Research Center. Joel Levine is heading up the program, which is being proposed for NASA's Scout-class missions. Levine's team first proposed the mission for the 2007 Mars landing opportunity.

"NASA received about thirty proposals, and after about 9 months they down-selected to four, and to our surprise and pleasure, ARES was one of the four selected for what they called 'step 2.' We each got $2 million – each of the four – and we were told to come up with a more detailed architecture and more detailed program plan, better budget, and so on. And that was due about 9 or 10 months after we were down-selected." The four mission "finalists" included ARES, the Phoenix polar lander, the Marvel orbiter (designed to study the composition of the atmosphere using high resolution spectrometers), and the SCIM, a mission that would orbit Mars about 50 miles above the surface, collecting a few quarts of Mars atmosphere to bring back to Earth.

Phoenix ultimately won out and became a highly successful polar mission. Levine's team was disappointed that their mission was not selected, but ARES wasn't finished, Levine relates. "An interesting thing happened. The NASA Aeronautics Program headquarters was so excited about the prospect of flying an airplane on Mars that when we were not selected, the associate administrator for aeronautics called up and said, 'Look, this is great for aeronautics – flying an airplane on Mars to study Mars, and we want to help you – to fund you – to develop the airplane so that the next opportunity you'll win. It will be a big feather in the cap for NASA Aeronautics.' So we got the $38 million in 2003 and we started a project here at NASA Langley called Planetary Airplane Risk Reduction Project – PARR. We had several dozen of our top aeronautical engineers doing wind tunnel tests and drop tests and computational fluid dynamics tests, and we really now understood how you deploy and fly an airplane in the atmosphere of Mars."

Levine's team spent 4 years honing the details of ARES and resubmitted it for the second round of Mars Scout proposals. ARES actually got higher grades than it had the first time around, but in the interim, the National Academy of Science pointed out that NASA had never attempted a mission to the upper atmosphere of Mars (called an aeronomy mission). So NASA decided that they wanted an upper atmosphere orbiter. "They basically decided what they wanted before the proposals came in," Levine laments. Still, he remains upbeat that ARES will be funded soon, perhaps at the next opportunity for Scout proposals. "We know a tremendous amount more than we did. Several years ago we did the first successful high altitude tests where we deployed a half-scale model of the Mars airplane at 103,000 ft, and everything worked perfectly."

Although orbiters are imaging the Martian surface in unprecedented detail, ARES has the advantage of in situ measurements in a critical area – the planetary boundary layer. It's one of the regions scientists know very little

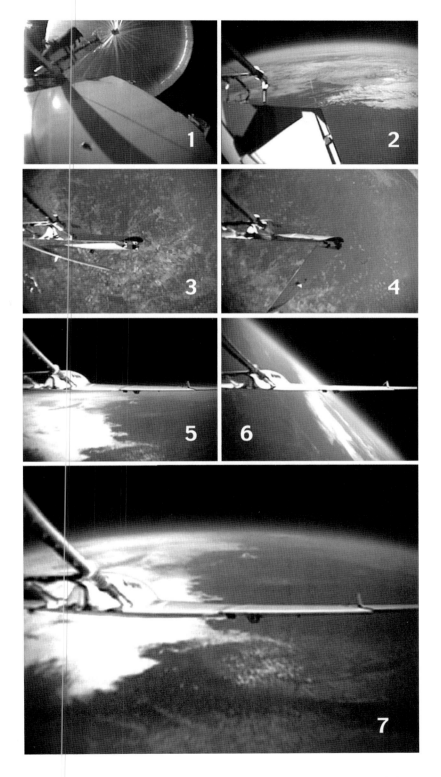

Video sequence of an ARES deploying wings at 103,000 ft altitude, where air pressure is similar to that on Mars at cruising altitude. 1 ascent on balloon; 2 tail deployed; 3 and 4 wings deployed; 5 ARES levels out; 6 aircraft makes a series of turns; 7 ARES in flight at 100,000 ft (ARES team/ NASA-Langley)

about. The boundary layer encompasses the first few kilometers of the Martian atmosphere. It is a very dynamic area where solar heating during the day causes the air to expand, while at night the air contracts. This breathing effect mixes

Langley's ARES team stands next to a full-scale model of their Mars craft (ARES team/ NASA-Langley)

chemistry of land and air. One of the problems engineers have in achieving pinpoint landings on Mars is a lack of understanding of the dynamics of the lower atmosphere, says veteran Mars scientist Ben Clark. "One of the interesting things from an atmospheric standpoint is that most, if not all, of the landers have landed a little further downrange than all the predictions. That means that there is something about the atmospheric density profile that we really don't understand yet."

ARES seeks to shed light on this mystery, Levine explains. "We will fly within the planetary boundary layer and we'll have a meteorological package. We'll measure the three dimensional structure of wind, pressure, temperature, and density as we fly. We don't have any of this information over long distances, so one of our experiments is to quantify the planetary boundary layer."

Because of the thin Martian air, ARES is controlled not by a propeller but by a rocket engine. To stay aloft, the craft will need to travel at speeds up to 450 miles per hour. With the fuel it can carry, the drone should be able to continue flying for 60–120 min. ARES is very light, with most materials consisting of non-metallic composites. The craft measures its environment once each minute, checking on atmospheric composition and structure. A tail-mounted camera sends real-time images of the landscape ahead, while a downward-looking camera images objects less than a few centimeters across.

ARES has a specific quarry in its sites – sources of methane. ESA's Mars Express and several ground-based observations have detected methane plumes in several areas on Mars, including Syrtis Major, a region of very ancient Noachian terrain. Researchers believe the methane has one of two sources: it is either related to volcanic activity, or to biogenic activity. Confirmation of either would be a landmark in planetary studies. And unlike a balloon, ARES is not slaved to where the wind blows, Levine says. "Our airplane has a navigation system. We decide where it goes and that's where it goes – not where the wind takes it. It's pre-programmed; there's an onboard computer. We tell it where we want it to go. For example, we have an idea where the methane is produced on Mars, and we tell it where to fly right over the methane, and what sort of maneuvers to do as it flies over the region of where the methane is produced. The methane, which is a gas produced biogenetically on Earth, appears to come from two or three places on Mars, so what we do is we pre-program it so the airplane will fly right over the methane. You can't do that with a balloon."

With its complement of atmospheric instruments and its good track record in tests, ARES shows great promise as a Mars explorer, Levine believes. "We have every expectation that ARES will fly in the near future."

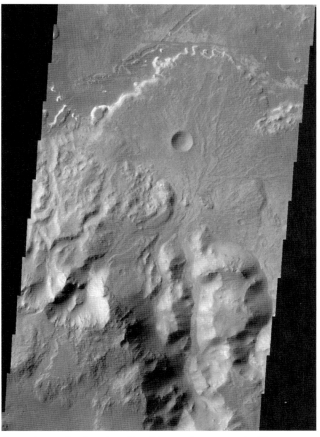

Several parcels of ancient Noachian real estate are leaking methane. This image covers a debris fan in Holden Crater (THEMIS image NASA/JPL/Arizona State University)

WHAT'S IN THE QUEUE?

Joel Levine's ARES Marsplane may be the next UAV to soar through planetary skies, but more conventional technologies are in line to fly, with confirmed launch dates scheduled. Orbiters and atmospheric probes make the list.[1]

First up, in 2011, is Japan's Venus Climate Orbiter, also known as Akatsuki. A Japanese press release emphatically states: "Akatsuki is the world's first planetary probe that deserves to be called a meteorological satellite." One of the mission's primary goals is to help scientists understand the hurricane-force superrotation of the Venusian atmosphere. Akatsuki will scrutinize the formation of the thick sulfuric acid clouds that envelop Venus, and search for lightning on the planet. Akatsuki's five cameras will map the movement of the Venusian atmosphere in three dimensions, by taking continuous images in different wavelengths ranging from infrared to ultraviolet. The multiple images will enable Akatsuki to obtain three-dimensional data.

1. Many missions take advantage of planetary flybys as gravity assists to slingshot them toward their ultimate destination. Although these missions typically gather information during these flybys, we will here explore only those specifically designed for the target planet.

Japan's Venus Climate Orbiter, called Akatsuki, will reveal Venus in three dimensions (© JAXA/Akihiro Ikeshita)

Although some Earth observation satellites have done so on Earth, no similar spacecraft has observed another planet in three dimensions. JPL planetary scientist Kevin Baines is excited about the mission. "It will be equatorial, and they've decided to move it in the direction of motion of the winds so they can loiter over the clouds and watch what goes on for about 10 h at a time. We will also be able to look for vertical waves that move from the surface through the atmosphere; some theories say that this surface coupling to the atmosphere, these waves, may have a lot to do with the superrotation. We can also measure winds at night, which no one has done yet. This enables us look at the thermal tide. The thermal tide acts like a sinusoid in these winds. Zonal winds should pick up and slow down and pick up again in a longitudinal pattern. You need to get data from all around the planet to do this."

Also in 2011, the Juno orbiter begins its odyssey toward Jupiter. While primarily a gravity mapping and magnetosphere mission, the polar orbiter will be able to record Jupiter's water abundances by use of microwave sounding. Planetary atmosphere authority Andy Ingersoll hopes Juno will be able to fill in some of those nasty blanks in our understanding of the formation of the outer planets. "Radio waves can penetrate the atmosphere to great depth, and the Juno spacecraft is going to measure the radio emissions coming up from the warm, deep atmosphere. They're hoping to sort out the water abundance from that."

The year 2013 will see the launch of MAVEN, the Mars Atmosphere and Volatile Evolution Mission.[2] Lockheed Martin's Mars expert Ben Clark is on the team of scientists who will monitor the craft as it samples the upper reaches of the Martian atmosphere. "Those kinds of missions have been proposed since the beginning of Mars exploration. They're called aeronomy missions for exploring the upper atmosphere. Upper atmosphere missions never were exciting enough that NASA would bite the bullet and do it. One of the things that [Principal Investigator] Bruce Jakosky did was to tie his mission to the theme of astrobiology. That is probably how we sold it." Clark says that researchers have to learn how Mars's atmosphere is being lost. Jakosky's team reasons that by measuring the losses in the atmosphere today, scientists can extrapolate backwards in time to see how much atmosphere has been lost over hundreds of millions (or billions) of years. It's critical information to know, Clark says. "If you don't make these measurements, you will remain in a great deal of ignorance. We designed that mission so that it comes down to a certain altitude as deep as you can afford to go. You

2. This is the mission that NASA selected over the ARES.

get much deeper and it will decay the orbit down and enter the atmosphere. We picked an altitude and instead of setting the altitude at so many kilometers, we set it at a certain pressure or density. The atmosphere is variable, and from a scientific standpoint what they're interested in is getting down to some density where the phenomena start changing. They want to measure everything above that. MAVEN is designed to systematically take lots and lots of measurements. For the most part, you have to have the whole set of measurements to understand what's going on: what ions are there, what their energies are, what directions they are moving in. Once you have the energy and direction of an ion, you can then know if it's going to escape the gravitational force and leave the planet. They'll be able to determine which gases are escaping and at what rates."

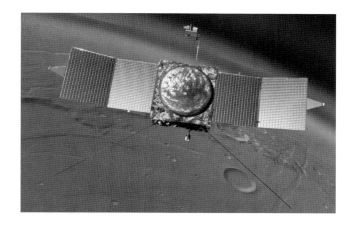

The MAVEN aeronomy mission will study the upper fringes of Mars's atmosphere (Spacecraft model NASA/JPL)

HOT OFF THE DRAWING BOARD…BUT NOT YET

Still other missions are vying for space on launchers. JPL's Viktor Kerzanovich has been working with a team to send an ambitious balloon mission, called VALOR, to the hothouse world of Venus. "We have been working for several years to fly a new type of balloon on Venus that will last for about 30 days, so we hope that this will be a very spectacular and very science-rich project. It's the VALOR program, a big balloon that can carry about 400 kilos above the Venus surface."

JPL's Kevin Baines believes that the next generation of science from Venus must come from within its environment. "You have to do what I call 'experiencing Venus' to get all the answers. We try to do chemistry from orbit, and we do, but it's all model-dependent on knowing the clouds. For example, if you have a high cloud with so many molecules of carbon monoxide, you assume that the cloud itself is made of it. But inside the cloud itself, you can get light rattling around and different things going on. When you try to derive the abundances of things, it can really be mucked up by the clouds. It also depends on lighting conditions, time of day. It's much better to be in situ, where you can float around and actually measure and count up the molecules accurately and not have to worry about modeling. Venus is a very chemically violent place, with so many different reactions going on compared to Earth's atmosphere. It's a laboratory for understanding all sorts of chemical atmospheric processes, and it does it all in one fell swoop in this big soup of an atmosphere. We really want to watch these chemical cycles going on in the atmosphere. It's very much temperature and pressure dependent as well, so as the balloon bobs up and down just a couple of kilometers, we expect the sulfur dioxide, for example, to change by almost a factor of ten.

We can use that information together with other chemical species to find out what the most important cycles on Venus are."

The key to VALOR's success will be longevity. The mission calls for a 30-day cruise in Venus's middle cloud deck. To that end, the system must be robust, explains Kerzhanovich. "The whole idea is to make it impermeable so it will not lose the gas in a month and maybe forever. It will be built of very strong material, the same material that (Mars Exploration Rover) air bags were made of. And remember on Mars when you dropped this 400 kilos of MER experiments with the speed of about 20 m/s they didn't pop."

Kevin Baines agrees that the mission is one of potentially great discovery. "It's like sailing the seas for the first time. I feel like Magellan. I don't know what's going to be on the other side. Who knows where we're going to end up? The balloon will go in at mid-latitudes. Originally, based on what we didn't know, we thought we might end up at the pole in a week or two. It might take the whole month to get there. They're now thinking the poleward motion depends on time of day, too. We really want to get to the poles because there's a different type of cloud regime there."

VALOR would use a balloon similar to this test version (JPL)

Another Venus mission under consideration is the SAGE (Surface and Atmospheric Geochemical Explorer). Its principal investigator, University of Colorado at Boulder's Larry Esposito, sees SAGE as a capable follow-on to his work on the Venus Express mission. "We're flying a very similar camera that would have better wavelength coverage. Just starting with the camera, it'll have better coverage in the ultraviolet that will allow us to see sulfur dioxide clearly, but the SAGE mission is going to land on the surface of Venus. So, that's a big difference from Venus Express and all the recent missions in the last 25 years."

As SAGE descends through the atmosphere, the probe would sample the air extensively. It would try to pin down some of those all-important noble gases, as well as refining estimates of other constituents, mapping the vertical profile of the atmosphere, and charting the winds. The probe would also have an aerial camera system to take images on the way down as well as on the surface. Once down, the craft would investigate the surface by irradiating it with lasers and neutrons. The lander would even be able to dig into the soil, Esposito says. "We have an arm that scrapes a hole in the surface and then we take pictures and shoot lasers into that hole so we can measure the gradient of the composition into the surface of Venus and potentially get below the top layer of the rocks, which have interacted with the atmosphere." SAGE is designed to last for 3 h in the hostile Venusian conditions, 2 h longer than any previous lander.

SAGE is in competition with two other proposals for NASA's next New Frontiers class mission. The first, called Osiris, would return a sample from an asteroid. The second is a lunar sample return mission called Moonrise. Says Esposito, "You can draw your own conclusions about whether it is easier to bring back a sample or go to the hellish Venus environment."

For nearly two decades, Mars was a prime target of study for balloon missions. One of the earliest and most well developed began as a student summer activity at Caltech some 20 years ago. Louis Friedman, an aerospace engineer and Executive Director of the Planetary Society, was a part of the study as it unfolded. "There was enough interest and innovation in what they were doing that nobody wanted to let it drop. At the same time, the Russians and the French had their heritage of Venus balloons and were pursuing this project. The Planetary Society had good relations with the Russians, so we ended up proposing to the Russians that we would

SAGE will scrutinize the Venusian atmosphere as it descends and deploys in stages, ultimately touching down in a geologically active area (NASA/JPL)

continue to fund this work and develop it with the idea that we could help the development of a Mars balloon for their 1994 Mars mission."

The Planetary Society sponsored a series of hands-on testing sessions using manned hot air balloons to fly concepts. They also launched balloons of different designs for Mars missions. But the real innovation came with the invention of a "snake" – a guide rope that dragged from the balloon probe at night. Friedman traces the design to conventional ballooning: "When you throw a guide rope out of a balloon, the balloon loses some of its mass and gains some buoyancy, so as the balloon begins to sag, you keep throwing out guide rope, and what you end up doing is allowing the

balloon to stay in the air longer. When the guide rope hits the ground, you basically become tethered to the ground because when you put more and more of it on the ground, you have less and less of it you're carrying, so you're going up higher, keeping a constant altitude. It's well known to balloonists; it's a standard technique."

The innovation, says Friedman, came later, when some of the engineers and scientists on the project proposed inserting instruments into the guide rope. "Some of the people including small spacecraft expert Jim Cantrell, said 'Why don't we instrument this guide rope?' Put actual instruments in it because they're going to be dragging on the ground and we can measure all kinds of chemical and surface properties and characteristics about the ground. So this became in instrumented guide rope, and we called it the snake because it slithered along the ground and looked like a snake."

That idea caught on. The Russians liked it and began to integrate it into the official project. Further testing was done. The instrumented snake

Early Mars balloon probes called for a pressurized helium balloon at the top with a Montgolfier solar-heated balloon below (© Michael Carroll/Planetary Society)

made the balloon more complicated than just a simple free flyer. And being an international project added yet another layer of complexity. But Friedman adds, "It was just the right kind of project for the Planetary Society because it involved students, contributed technical innovation, got out a flight project, ended up helping the Space Agency, and pushed the whole thing along." Sadly, after numerous postponements the balloon portion of the mission was canceled, and the ultimate project, Mars 96, failed during launch.

Are there balloons in future Mars exploration? JPL's Viktor Kerzhanovich, veteran of both the Soviet and U.S. space programs, is skeptical. Kerzhanovich believes that – in light of advances in orbital reconnaissance – the time has passed for Mars balloon probes. "For 10 years or more we pursued the Mars balloon technology and believe we have really good success on that, but the problem with Mars balloons is that there is no need for them now.

After the Mars Reconnaissance Orbiter, the upcoming Mars Science Laboratory, and the Mars Exploration Rovers, there is no scientific task now for this balloon. It was OK in 1987 or up to the 1990s, but there is no great scientific demand now."

MRO imaging scientist Steve Lee disagrees. "You've got these suites of instruments that you can bring to bear on the problem; in the earlier missions you were much more limited. I would think that balloons would still play a part because you're sort of in situ to do the measurements. You're at a certain elevation, and it would give you detailed information on the winds because you're in them as opposed to trying to interpret what the winds are by looking at something drifting from one orbit to the next."

Veteran JPL mission designer Charley Kohlhase suggests another Mars mission that is worthy of scarce planetary exploration resources: "The Holy Grail is a Mars sample return mission. Ever since I was a graduate student, that's always been 15 years in the future. Thirty years later we're

Later concepts of the Blamont/Planetary Society study envisioned a balloon that would cruise at 10,000 ft during the day (left), then settle to the surface at night. Night sessions would utilize instruments in the "Snake" to measure soil composition (© Michael Carroll/Planetary Society)

no closer to doing it than we were back in the 1980s. I don't do the cosmo-chemistry, petrology, geology, or geo-chronology things that they'd be able to do, but I think that having a sample in hand that you could throw at all the world's laboratories – you know, the combined expertise of everybody – is going to go a long way towards figuring out sort of the time scale of things on Mars."

A sample return mission is important not only in the study of geology but also in the study of atmosphere and climate. One mission is leading up to the kind of automation needed in a sample return. It's a Russian mission called Phobos Grunt ("grunt" is Russian for "soil"). This ambitious Mars orbiter is designed to study Mars from orbit while catching up to Mars's larger moon, Phobos. A lander will take in soil and rock, then launch again to return Phobos samples to Earth. Phobos Grunt is currently scheduled for a 2011 launch. Says Lou Friedman, "The only country on Earth that's done a successful automatic surface sample return mission is the Soviet Union. And the Japanese are now trying it with Hyabusa (asteroid sample return mission), and I sure hope they make it. But it's something the U.S. hasn't done yet, and I think in many aspects the upcoming Phobos Grunt mission is seen as very important, as we build up the capability for a Mars sample return. And I think that's becoming understood finally as this mission gets more serious. It's all a test whether the Russians are going to be back in the planetary science business."

COOL CRAFT

Other studies are under way for probes even farther afield. For outer planet missions, engineers are considering a Jupiter Deep Probe. The craft would be smaller than the Galileo atmospheric probe. While Galileo reached 22 bars of pressure, the Jupiter Deep Probe's goal is 100 bars. To accomplish this goal, the craft would spend most of its descent without a parachute, free-falling for hundreds of miles. The challenge is to get the data up through the dense atmosphere to a relay craft hurtling by above, and to do it in time for the relay craft to receive and transmit the data back to Earth.

Beyond Jupiter, excitement is building for a balloon mission or advanced entry probe for Saturn's moon Titan. In 2008, NASA had to choose between funding for a flagship mission to Europa, Jupiter's oceanic moon, or a NASA/ESA

NASA study of a Montgolfier-style balloon in Titan's atmosphere (NASA/JPL/Wikipedia Commons)

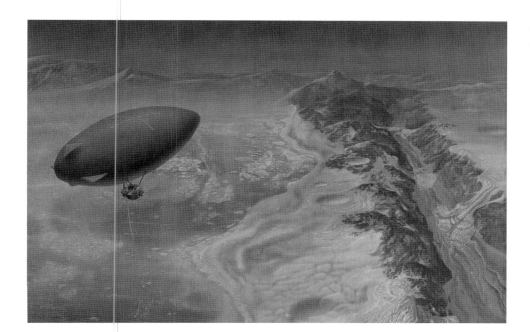

SAIC study of a Titan blimp probe, ca. 1982 (Art © Michael Carroll)

Titan orbiter with a Montgolfier-style balloon probe. The competition was stiff, but Europa won out. The Titan balloon probe was studied in detail and may still be funded sooner for a flagship mission (in about 2018). In a 2000 paper, Titan expert and team member Ralph Lorenz outlined reasons for deploying a Montgolfier Titan balloon system. "The potential for prebiotic materials at various locations and the need to monitor Titan's meteorology favor future missions that may exploit Titan's unique thick-atmosphere, low-gravity environment…a mobile platform like an airship or helicopter, able to explore on global scales, but access the surface for in situ chemical analysis and probe the interior by electromagnetic and seismic means. Such missions have dramatic potential to capture the public's imagination, on both sides of the Atlantic."

The ESA/NASA Titan study included a nuclear source to heat the open-air balloon. Other studies envision a Titan blimp or dirigible. As far back as 1982, the Science Applications Institute Corporation devised a plan for a Titan blimp with a cross-shaped antenna embedded in the top fabric. A more recent JPL study proposes a blimp that could drop canisters to retrieve samples of soil or liquid from Titan's hydrocarbon lakes.

Balloons aren't the only design under consideration, adds Lockheed Martin's Ben Clark. "We're working on a Titan mission now to land in one of the lakes. The lakes must be made of methane, which is the primary volatile in Titan's atmosphere other than nitrogen. There may be some other compounds such as ethane, a little propane, maybe some others. The real question is are these lakes shallow, and just represent some runoff from methane that rains out in the polar areas, or are there 'methanifers' akin to Earth's aquifers. Are there large-scale reservoirs of liquid methane near the

An advanced aerobot drops a sample-retrieval canister onto the lakeshore of one of Titan's polar methane seas (© Michael Carroll)

surface that feed these lakes?" Clark's probe would measure the depth and composition of the lakes, comparing their composition with that of the atmosphere to see if the liquid bodies are purely due to condensation. With that set of data, experts say they'll be able to provide an answer or constrain their models. Titan's sky continually precipitates a fine hydrocarbon haze, similar to fine particles of soot, Clark explains. "You have some turbulence in the atmosphere, and you have this low gravity, which leads to a very slow sifting rate. Eventually, some of this stuff ends up in the lake, but we don't know if it settles to the bottom or if turbulence in the lake itself keeps it stirred up. All that stuff is generated by the atmosphere; it's basically smog." The bodies of liquid on Titan are so closely tied to multiple cycles in Titan's atmosphere that studying Titan's lakes is tantamount to studying its atmosphere.

Astronomer Heidi Hammel bemoans the fact that no missions to Uranus or Neptune have yet made the cut. "Uranus and Neptune so far haven't been a top priority for planetary exploration because they are so hard to get to. It takes 10 years to get there, and the federal budget doesn't work on decadal timescales. They like missions to Mars; you plan them and launch them and have the data back in 2 years. But you never give up hope."

Hammel might have reason for a glimmer of hope: NASA's Goddard Space Flight Center recently undertook a feasibility study of a mission to Neptune, with a probe to its geyser-laden moon Triton. The Neptune Triton Explorer (NExTEP) would use solar electric propulsion and aerocapture to reach its final orbit around the blue giant. The Neptune orbiter would carry a Triton lander and at least one Neptune atmospheric probe designed to survive to the 200 bar depth. A 2003 Solar System Exploration Decadal Survey listed a Neptune orbiter and Triton Explorer as a "deferred high-priority flight mission." This sets the mission up for follow-on studies that might lead to its consideration in 2020–2030.

Forty years ago, engineers dreamed of sending probes to the skies of Venus or Titan. With advances in miniaturization, electronics, and materials, those dreams became reality. The bizarre designs of early visionaries morphed into practical machines, robots to sail the high frontiers of distant worlds. What will tomorrow's advances bring?

Future human outposts may include inhabited meteorological stations – or even colonies – in the atmosphere of Venus, as projected by this artist's interpretation of a NASA/ Glenn Research Center study. Breathable gases are lifting gases on Venus, so the inflated habitats double as buoyancy envelopes. Note the wind turbines for power generation and the bathysphere lowering a crew to greater depths, foreground (© Michael Carroll)

Chapter 11

To Venture on Wilder Seas

In 1577, on the occasion of his departure from Portsmouth on the *Golden Hind*, Sir Francis Drake offered the following prayer:

Disturb us, Lord,
When we are too pleased with ourselves,
When our dreams have come true
Because we dreamed too little,
When we arrived safely
Because we sailed too close to the shore.
Disturb us, Lord,
When with the abundance of things we possess
We have lost our thirst
For the waters of life;
Having fallen in love with life,
We have ceased to dream of eternity
And in our efforts to build a new earth,
We have allowed our vision
Of the new Heaven to dim.
Disturb us, Lord, to dare more boldly,
To venture on wilder seas
Where storms will show Your mastery;
Where losing sight of land,
We shall find the stars.
We ask you to push back
The horizons of our hopes;
And to push back the future
In strength, courage, hope, and love.

We continue to dream of sailing wilder seas to ever-more distant shores and the skies above them. No longer do our robot explorers skirt our terrestrial harbor, but rather, venture to the outer reaches of our Solar System. History shows us that the natural outgrowth of robotic exploration is often human adventure. The Wood's Hole research vessel *Knorr* discovered the wreck of the Titanic, and Robert Ballard's team sent robots to journey across its decks, down its staircases, and over its debris fields. But once that was done, tourists ventured into the abyss to witness the site of the historic sunken ship. The Rangers, Lunas, and Surveyors stood as precursors to human lunar voyagers. Even the eighteenth-century Montgolfiers, with their crew of farmyard animals, set their sights on going up in a balloon themselves.

Aerospace engineer and visionary Robert Zubrin sees this progression as a biological paradigm. He asserts that, simply put, a species that fails to expand dies. Aside from any biological imperative, human curiosity – its drive to discover what is "around the next corner" – is responsible for history's greatest inventions, most important technological advances, and most inspiring voyages. And so it follows that our robotic explorations of alien

M. Carroll, *Drifting on Alien Winds: Exploring the Skies and Weather of Other Worlds*,
DOI 10.1007/978-1-4419-6917-0_11, © Springer Science+Business Media, LLC 2011

skies will ultimately result in human travels there. Human experiential power trumps even the most finely automated emissaries.

Will humans take to the alien winds of Mars, Venus, or Saturn? What shape might these advanced human expeditions take? Form follows function, so designs of any advanced vessels will be driven by the environments for which they are destined. Steamy Venus poses extreme challenges for thermal control. The cryogenic atmospheres of the outer planets, on the other end of the temperature scale, offer their own difficulties. Geoffrey Landis, a researcher at NASA's Glenn Research Center, recently completed a NASA-funded study to see what it would take to set up camp in the clouds of Venus. The floating outpost would cruise at an altitude of about 50 km, where pressure is about 1 bar and the air is room temperature, a comparatively benign environment. Landis points out that Venus is rich in resources necessary for life – carbon, hydrogen, oxygen, nitrogen, and sulfur. Its dense atmosphere provides protection from cosmic radiation, in contrast to environments on Mars, the Moon, or asteroids. Gravity is 90% of that on Earth, so the long-term physical issues that may arise on Mars or the Moon are not a serious issue on Venus. Additionally, solar energy is abundant above the cloud tops. An added benefit of a Venusian location is that because of the density of Venus' CO_2 atmosphere, a breathable atmosphere serves as a lifting gas, about half as buoyant as helium is in Earth's atmosphere. A 400-m radius envelope – about the size of a small sports arena – can lift 350,000 tons, a mass equivalent to the Empire State Building.

Because of the global winds in the middle cloud deck of Venus, the aerostat outpost would not remain stationary but would continually circle the globe. New ground could continually be studied, and relay stations could be deployed to keep in contact with surface telerobotics. Sulfuric acid droplets would constantly batter the station, but engineers have come up with plenty of acid-resistant technologies that could be brought to bear. Landis's report concludes, "In short, the atmosphere of Venus is the most earthlike environment in the Solar System. Although humans cannot breathe the atmosphere, pressure vessels are not required to maintain one atmosphere of habitat pressure, and pressure suits are not required for humans outside the habitat…in the near term, human exploration of Venus could take place from aerostat vehicles in the atmosphere… in the long term, permanent settlements could be made in the form of cities designed to float at about 50 km altitude in the atmosphere of Venus."

"It's a fun subject," says the Planetary Society's Louis Friedman. "Several people, including [astronomer Carl] Sagan, pointed out that the conditions at the 50 km altitude in Venus are rather clement compared to the surface, and even compared to Earth: reasonable pressure, reasonable atmospheric density. There are a few disadvantages, such as sulfuric acid clouds and the excess carbon dioxide, but still it's rather believable as a place where you could float habitats. I still regard it as science fiction, but that's only because I'm not an advanced thinker. My planning horizon only goes out a couple of 100 years."

For Larry Esposito of the Laboratory for Atmospheric and Space Physics, the jump to human exploration is a bit premature but one that humanity is moving toward. "The first step toward inhabiting Venus or Mars is to learn more about the environment. To float around on Venus, I would definitely go in order to carry out scientific investigations. Just for the view, I'd rather go to the Mediterranean."

To Lou Friedman, Mars is still the best bet for future settlement. "Mars still stands out. Mars keeps getting better and better and better. And the combination of discoveries on Earth and at Mars that made Mars a possible place where life once existed ‒‒ and certainly a place where life will exist in the future – is what has motivated me the most."

Whether or not it is the destiny of humans to drift upon alien winds, one thing is certain: humankind has benefited immensely from the study of other worlds. Friedman asserts, "The greatest single benefit of space science and space exploration is in the understanding about our own planet ‒‒ the understanding of Earth. Discoveries have led to deep insights we have about global climate change and the insights we have about dust transit and the transition of storms in the atmosphere. I think what we've learned of the planets has basically been an enormous step forward in the understanding of our own planet."

Piloted dirigibles might be possible, even in the thin Martian air. Here, an MIT study envisions an airship with a black gas envelope to utilize solar heating of the gases inside. Note the solar cells on top for power, and the large turbines to move the craft in Mars' 7 mbar pressure (© Michael Carroll)

Lockheed/Martin's Ben Clarke sees a more fundamental understanding from space exploration and, more specifically, from atmospheric research, that of "understanding where we came from. That's kind of a biological question, but you can't have the biology unless you have the suitable environment. A suitable environment involves more about atmospheres than probably anything. You want liquid water, but you can't have liquid water unless you have the right temperature and pressure. You probably cannot have life on a comet or an asteroid, because you don't have gas. This is something that very few people appreciate; it's seldom talked about. Most biological organisms need more than just liquid or solid nutrients. They need gas. One reason is to keep water in a liquid phase; it boils in a vacuum. The other reason is that, for example, plants take up CO_2. Nitrogen gets fixed to make nitrogen nutrients. Sulfur is used; in fact, sulfur was one of the earliest ways to do photosynthesis. Oxygen is taken up by organisms that respire, which includes all animals, some plants, and a lot of bacteria. Atmospheric gases play a very major role in life."

Phoenix lander Principal Investigator Peter Smith adds that space exploration is an investment. "Knowledge is the payoff. People often worry about the cost. They wonder why we should spend money to go to space when we have problems on the Earth like hunger and health problems and other things, but even when governments spend huge amounts of money – hundreds of billions of dollars – they don't solve these problems. The idea that you would take an extra ½ of 1% of the federal budget and apply it to one of these huge problems is not going to make a difference, so you've lost your space program and gained essentially nothing. You don't send the money into space. The money is spent employing our best and our brightest engineers, scientists, and aerospace people; it keeps us on our toes and on the front edge of technology and science. It gives our country a brain trust that is invaluable. Without any of these programs, our scientists would be working on how to build better bombs, which I don't think is something we should be very proud of. I think when we look back in history, the last 50 years of the space program have been hugely productive. One of the great gifts we give to the rest of the world as Americans is that we offer the data from space freely to all scientists and interested parties around the world. You see pictures of the planets and moons used in so many different ways, and always with great pride. This is our Solar System that we're looking at here. It's not 'The Americans have given us a few lousy pictures.' It's more like 'These are great! Keep doing it!' It's one way that America really is an ambassador to the world. Everybody appreciates good solid data about the worlds around us."

Still other researchers are attracted to the siren call of the outer planets. Cassini's Carolyn Porco envisions the sites at the gas giants. "We see all this fabulous structure and the ornate atmospheric motions in Jupiter and Saturn, but if you were a human being in a tiny little spacecraft, is the atmospheric scale invariant? Do we see all that structure also down at the small scale? If you put me in the atmosphere of Jupiter, the first thing that I would be impressed by is that whereas you can see the curvature of Earth here, the thing would look like it had no curvature because it's so bloody big. It would look like you were seeing clouds to infinity. On our horizon, you can still make out cloud structures because they are only tens of miles away. But on Jupiter they could be thousands of miles away, so you would see clouds further and further away, but they would become featureless. I don't know if you would get a sense of being in the Great Red Spot, for example. But you would be able to see colors that we don't see, so that would be nice."

Caltech's Andy Ingersoll has another destination in mind. "I think I would pick Saturn, just because of the rings. The view of the sky with those rings up there would just have to be mind-blowing. I'd pick some place at mid-latitude in the summer hemisphere because that side is illuminated. Especially at night it would be a spectacular site to look out and see the rings from one horizon to the other."

For now, our cosmic wanderlust must be fed by remote sensing and our robotic emissaries. These sources provide immense inspiration. Andy Ingersoll speaks fondly of the Voyager missions and their effect on human inspiration. "For me, I keep thinking, 'Gee, the planets have come through again!' They are our heroes. They keep coming up with surprises. The volcanoes of Io, just because it's so far away from the Sun and should be a dead world but its so active. It was an enormous surprise. The other was the activity of Neptune. There we were in the outer Solar System where the temperatures were 60 or 70 K, and yet the winds are very strong and you had a whole interesting weather system, so with my own specialty as a meteorologist, Neptune was an enormous surprise. But I could go on. In fact, it's the number of surprises that's really impressive."

That number of surprises, and the inspiration gained from the planets and moons around us, will undoubtedly rise as we explore the skies of distant worlds and drift on their alien winds.

Appendix 1: Venus Missions

Sputnik 7 (USSR)	Launch 02/04/1961	First attempted Venus atmosphere craft; upper stage failed to leave Earth orbit
Venera 1 (USSR)	Launch 02/12/1961	First attempted flyby; contact lost en route
Mariner 1 (US)	Launch 07/22/1961	Attempted flyby; launch failure
Sputnik 19 (USSR)	Launch 08/25/1962	Attempted flyby, stranded in Earth orbit
Mariner 2 (US)	Launch 08/27/1962	First successful Venus flyby
Sputnik 20 (USSR)	Launch 09/01/1962	Attempted flyby, upper stage failure
Sputnik 21 (USSR)	Launch 09/12/1962	Attempted flyby, upper stage failure
Cosmos 21 (USSR)	Launch 11/11/1963	Possible Venera engineering test flight or attempted flyby
Venera 1964A (USSR)	Launch 02/19/1964	Attempted flyby, launch failure
Venera 1964B (USSR)	Launch 03/01/1964	Attempted flyby, launch failure
Cosmos 27 (USSR)	Launch 03/27/1964	Attempted flyby, upper stage failure
Zond 1 (USSR)	Launch 04/02/1964	Venus flyby, contact lost May 14; flyby July 14
Venera 2 (USSR)	Launch 11/12/1965	Venus flyby, contact lost en route
Venera 3 (USSR)	Launch 11/16/1965	Venus lander, contact lost en route, first Venus impact March 1, 1966
Cosmos 96 (USSR)	Launch 11/23/1965	Possible attempted landing, craft fragmented in Earth orbit
Venera 1965A (USSR)	Launch 11/23/1965	Flyby attempt (launch failure)
Venera 4 (USSR)	Launch 06/12/1967	Successful atmospheric probe, arrived at Venus 10/18/1967
Mariner 5 (US)	Launch 06/14/1967	Successful flyby 10/19/1967
Cosmos 167 (USSR)	Launch 06/17/1967	Attempted atmospheric probe, stranded in Earth orbit
Venera 5 (USSR)	Launch 01/05/1969	Returned atmospheric data for 53 min on 05/16/1969

M. Carroll, *Drifting on Alien Winds: Exploring the Skies and Weather of Other Worlds*,
DOI 10.1007/978-1-4419-6917-0, © Springer Science+Business Media, LLC 2011

Venera 6 (USSR)	Launch 01/10/1969	Returned atmospheric data for 51 min on 05/17/1969
Venera 7 (USSR)	Launch 08/17, 1970	Lander returned data during descent and for 23 min on surface
Cosmos 359 (USSR)	Launch 08/22/1970	Possible lander; upper stage failure
Venera 8 (USSR)	Launch 03/27/1972	Atmospheric probe and lander; transmitted from surface for 50 min
Cosmos 482 (USSR)	Launch 03/31/1972	Upper stage failure
Mariner 10 (US)	Launch 11/04/1973	Venus flyby en route to Mercury; first closeup images of Venus
Venera 9 (USSR)	Launch 06/08/1975	Orbiter/lander; atmospheric studies, first images from surface of another planet, first artificial satellite of Venus 10/22
Venera 10 (USSR)	Launch 06/14/1975	Orbiter/lander; atmospheric studies, images from surface
Pioneer Venus 1 (US)	Launch 05/20/1978	Radar orbiter, atmosphere studies
Pioneer Venus 2 (US)	Launch 08/08/1978	Bus and atmospheric probes (4)
Venera 11 (USSR)	Launch 09/09/1978	Orbiter/lander; some instruments failed, no imagery, detection of possible lightning
Venera 12 (USSR)	Launch 09/14/1978	Orbiter/lander; some instruments failed, no imagery
Venera 13 (USSR)	Launch 10/30/1981	Orbiter/lander; atmospheric studies and surface panoramas
Venera 14 (USSR)	Launch 11/04/1981	Orbiter/lander; atmospheric studies and surface panoramas
Venera 15 (USSR)	Launch 06/02/1983	Orbiter with radar mapper
Venera 16 (USSR)	Launch 06/07/1983	Orbiter with radar mapper
Vega 1 (USSR)	Launch 12/15/1984	Lander with balloon/Comet Halley flyby/first use of balloon on another world
Vega 2 (USSR)	Launch 12/21/1984	Lander with balloon/Comet Halley flyby
Magellan (US)	Launch 05/04/1989	Radar mapping orbiter
Galileo (US)	Launch 10/18/1989	Venus flyby en route to Jupiter
Cassini (US/ESA)	Launch 10/15/1997	Venus flyby en route to Saturn
MESSENGER (US)	Launch 08/03/2004	Two Venus flybys en route to Mercury orbit
Venus Express (ESA)	Launch 11/09/2005	Second European planetary spacecraft; Venus orbiter; first images of Venus south pole

Appendix 2: Mars Missions

Mars1960A (USSR)	Launch 10/10/1960	Possible Mars flyby attempt; launch failure
Mars1960B (USSR)	Launch 10/14/1960	Mars flyby attempt; launch failure
Sputnik 22 (USSR)	Launch 10/24/1962	Mars flyby attempt; launch failure
Mars 1 (USSR)	Launch 11/01/1962	Mars flyby attempt; interplanetary studies carried out; contact lost 03/21/1963
Sputnik 24 (USSR)	Launch 11/04/1962	Attempted Mars landing; upper stage failure; reentry 01/19/1963
Mariner 3 (US)	Launch 11/05/1964	Attempted Mars flyby; shroud failure
Mariner 4 (US)	Launch 11/28/1964	First successful Mars flyby 07/15 and 16, 1965; 22 photos and other data received
Zond 2 (USSR)	Launch 11/30/1964	Flyby/lander attempt; contact lost May of 1965
Zond 3 (USSR)	Launch 07/20/1965	Lunar flyby; returned images of Moon's far side, flew test Mars trajectory
Mariner 6 (US)	Launch 02/26/1969	Second successful Mars flyby 07/31; 75 images and other data returned
Mariner 7 (US)	Launch 03/27/1969	Third Mars flyby; 126 images of southern hemisphere
Mars1969A (USSR)	Launch 03/27/1969	Flyby attempt; launch failure
Mars1969B (USSR)	Launch 04/02/1969	Flyby attempt; launch failure
Mariner 8 (US)	Launch 05/08/1971	Flyby attempt; launch failure
Cosmos 419 (USSR)	Launch 05/10/1971	Attempted orbiter/lander; upper stage failure
Mariner 9 (US)	Launch 05/30/1971	First orbiter of another planet: Mars orbit achieved November 14; global mapping continued until 10/27/1972
Mars 2 (USSR)	Launch 05/19/1971	Successful Mars orbiter (achieved November 27); lander failed

Mars 3 (USSR)	Launch 05/28/1971	Successful orbiter (December 2); lander touched down 12/02/1971 but ceased transmission after just 20 s. A tethered rover was also aboard
Mars 4 (USSR)	Launch 07/21/1973	Attempted orbit, retros failed; flyby on 02/10/1974 returned some data
Mars 5 (USSR)	Launch 07/25/1973	Achieved orbit 0212/1974; transmitter failure after 22 orbits
Mars 6 (USSR)	Launch 08/05/1973	Lander attempt; contact lost at landing; first in situ measurements of the Martian atmosphere
Mars 7 (USSR)	Launch 08/09/1973	Attempted lander; premature separation caused craft to miss planet; flyby on 03/09/1974
Viking 1 (US)	Launch 08/20/1975	Orbiter global mapping/first successful Mars landing 07/20/1976; operated on surface for over 5 years
Viking 2 (US)	Launch 09/09/1975	Orbiter global mapping/successful Mars landing 09/03/1976; surface operations for 3½ years
Phobos 1 (USSR)	Launch 07/07/1988	Attempted orbiter and two Phobos landings; contact lost en route
Phobos 2 (USSR)	Launch 07/12/1988	Mars orbiter; attempted Phobos landings; orbiter failed just prior to final encounter with Phobos before landers could be deployed
Mars Observer (US)	Launch 09/25/1992	Mars Orbit attempt; contact lost prior to orbit insertion, probably due to a ruptured fuel line
Mars Global Surveyor (US)	Launch 11/07/1996	Mars orbiter; first use of aerobraking for planetary orbit; lander relay
Mars 96 (USSR)	Launch 11/16/1996	Attempted orbiter and multiple landers; launch failure
Mars Pathfinder (US)	Launch 12/04/1996	Airbag equipped lander and rover; mission lasted 3 month
Nozomi (Japan)	Launch 07/03/1998	Orbit attempt, abandoned due to fuel loss; Mars flyby 12/14/2003
Mars Climate Orbiter (US)	Launch 12/11/1998	Orbit attempt; spacecraft burned up in Martian atmosphere due to a navigational error
Mars Polar Lander/ Deep Space 2 (US)	Launch 01/03/1999	Attempted lander and penetrators; crash probably due to premature engine cutoff from software problem
Mars Odyssey (US)	Launch 03/07/2001	Orbiter reached Mars 10/23/2001; lander relay

Mars Express (ESA)	Launch 06/02/2003	Orbiter and attempted lander Beagle 2; orbit achieved 12/25; international lander relay; communications with Beagle lander lost before landing
Spirit (MER-A) (US)	Launch 06/10/2003	Mars rover, airbag landing; has explored over 7.7 km of Gusev Crater/Columbia Hills since landing 01/03/2004
Opportunity (MER-B) (US)	Launch 07/07/2003	Mars rover, airbag landing; has explored over 26.6 km of the Meridiani plains since landing 01/25/2004
Mars Reconnaissance Orbiter (US)	Launch 08/10/2005	Mars orbit 03/10/2006; high resolution imagery for future landing sites; search for evidence of present and past water
Phoenix	Launch 08/04/2007	Landed near north pole; mission from 05/25/2007–11/02/2007

Appendix 3: Outer Planet Missions

Pioneer 10 (US)	Launch 03/03/1972	First outer planets mission, first Jupiter flyby 12/04/1973
Pioneer 11 (US)	Launch 04/05/1973	Jupiter flyby, first Saturn flyby 09/01/1979
Voyager 1 (US)	Launch 09/05/1975	Jupiter/Saturn flyby
Voyager 2 (US)	Launch 08/20/1975	"Grand tour" flybys of Jupiter, Saturn, Uranus and Neptune
Galileo (US)	Launch 10/18/1989	Achieved Jupiter orbit 12/7/1995, deployed atmospheric probe
Ulysses (ESA)	Launch 10/6/1990	1992 and 1994 Jupiter flybys en route to study Sun's polar regions
Cassini/Huygens (US/ESA)	Launch 10/15/1997	Saturn orbiter/Titan probe
New Horizons (US)	Launch 1/19/2006	Jupiter flyby 2/28/2007 en route to Pluto encounter 7/14/2015

Chapter Notes and Sources

Chapter One

Legend has it that Spanish ships stranded while transporting horses to the West Indies would throw their horses overboard in an effort to save water rations. Another version has it that the Horse Latitudes refer to the "dead horse" ritual, in which sailors celebrate the point in their journey when they have worked off the debt of their advanced payment. Deck hands paraded a straw horse dummy, and then threw it overboard.

The Earth is nearly identical, in size, to Venus. Considering the diversity of our Solar System – where Pluto is the size of Earth's Moon and a thousand Earths would fit inside Jupiter – Venus and Earth are practically twins.

Atmospheric friction causes the tiny particles of rock or dust to vaporize as they enter the atmosphere. Earth's gravity pulls in debris at speeds from 7 to 45 miles/s (depending on the relative speed and direction of the incoming meteor). At this speed, any object hitting the atmosphere experiences severe entry heat from the friction of air molecules against it. Only large meteors survive long enough to slow down and impact the surface. According to work done by Jay Melosh and others at Arizona's Lunar and Planetary Laboratory, "Typical impact velocities are 17 km/s for asteroids and 51 km/s for comets."

The Mayan codex of Dresden dedicates a great deal of text to the rain god Chaac and to monsoon seasons and floods. Mayan codices, or books, are arranged in folds much like an accordion. Few survive today. The Dresden Codex is dated at the eleventh or twelfth century, but was copied from an original that may date back 400 years earlier. For more, see *Ancient Astronomy* by Clive Ruggles (published by ABC-CLIO, 2005).

It is estimated that between 37,000 and 78,000 tons of meteoritic material reaches the surface of the Earth each year... Most of this material falls as microscopic dust particles. A 1996 study estimated that between 36 and 166 meteors heavier than 10 g hit each million square kilometers of Earth's surface each year (up to 84,000 of that size annually). The study, by P.L. Bland, appeared in the *Monthly Notices of the Royal Astronomical Society*.

Only one thing enables Earth to retain free-floating oxygen in such quantities: life. Some studies estimate that without the photosynthesis of plant life, Earth would run out of oxygen within 5,000 years.

This vast pinwheel of violence can span 300 miles across, with winds up to 185 mph. When they form in the northern hemisphere, these great storms spin in a counterclockwise motion, with rising air at their center. In the southern hemisphere, they rotate clockwise with rising air masses at the center. Hurricanes, typhoons, and cyclones are all cyclonic storms – they rotate.

Chapter Two

To the Hebrews and early Christians, it was the "first heaven," the place where rains came from and winds were born. The "second heaven" was the realm in which the Sun, Moon, and stars resided. The "third heaven" was the place where God lived.

Leonardo began to carefully organize his thoughts of flight into his sketchbooks. The artist also understood a few key concepts about atmosphere. In his sixteenth-century book on painting, Leonardo pointed out that sunlit air adds blue light to a scene. He noticed that the longer the path through the air, the more blue light is added. Leonardo was describing Rayleigh scattering through his art some three centuries before Lord Rayleigh documented the effect in scientific literature.

Leonardo detailed one machine, in particular, which he called "the great bird." Ever cautious, Leonardo suggested that early flights be carried out over water for safe crash-landings. He even designed a flotation cushion under the pilot's platform to keep his invention from sinking.

Before them, bobbing in an autumn breeze, floated a bright blue varnished taffeta bag filled with hot air. The first Mongolfier balloon, uninhabited, consisted of a spherical envelope made of sackcloth lined with paper. Its sections were held together by nearly 2,000 buttons. An outer web of netting reinforced the entire balloon.

Its crew consisted of a sheep, a duck, and a rooster. The sheep's name was Montauciel, French for "climb to the sky."

The brother inventors certainly were, and set to work on a balloon that eventually carried the first humans into the skies above the French countryside. The Montgolfiers began to give rides to human passengers on tethered flights, but on November 21, a new and larger balloon took two men into the air in free flight. Jean-Francois Pilatre de Rozier and Francois Laurnet, Marquis d'Arlandes, joined the creatures of the air for 25 min.

*Called "fire arrows" by their first-millennium Chinese inventors…*Gunpowder was probably invented in China around the first century AD, but the first historical record of the use of rockets was in the battle of Kai-Keng between the Chinese and Mongols in 1232 AD. The Chinese capped a tube (probably bamboo) and left the other end open, packing it with gunpowder and a fuse. The tube was attached to a long stick that acted as a primitive guidance system, similar to a modern bottle rocket.

Dark patterns called absorption lines appeared in unique patterns across the spectrum, revealing what materials were present in the objects being studied. These lines are now known as Fraunhofer lines, named after Joseph von Fraunhofer, who first documented 570 of them.

To get around this problem, all planetary explorers have been covered in ablating materials that gradually burn off during entry. The preferred material is carbon fiber. To date, it has been used in all planetary missions with great success.

Chapter Three

Mariner 4 ... took 21 full images of the Martian surface, the first images returned from deep space. The spacecraft also returned 22 lines of a 22nd image, and shot three images past the terminator (on the night side of Mars).

The galactic ghoul was at it again 7 years later, when the Russian Space Agency attempted one of the most ambitious international flights to Mars. By this time, the Soviet Union had dissolved. Mars 96 used resources left over from the waning Soviet era. The failure of Mars 96 left what remained of the Soviet planetary program in shambles. But the program would recover. Even now, plans are progressing for a Russian return to Mars with the Phobos Grunt mission.

The Mars 96 mission consisted of a massive orbiter, two landers, and two penetrator probes. Penetrators are designed to impale the surface of a planet or moon without slowing down. The spear-like probes are ideal for seismic or other studies requiring a firm contact with the ground.

Nothing in the Asteroid Belt is big enough to be considered a planet. Several asteroids are large enough to retain a spherical shape and were once called "minor planets." Evidence suggests that Vesta has undergone chemical differentiation of some kind. Nevertheless, if all asteroids in the main belt were combined, their mass would be less than Earth's Moon. This is somewhat baffling because of their makeup. Some asteroids are carbonaceous, apparently coming from the ancient solar nebula from which all planets came. Their presence makes sense if the asteroids never assembled themselves into a planet. But others contain materials such as olivine, a common volcanic glass that is likely formed within the heart of a large planet. The origin of the asteroids, as well as their evolution, remain mysteries.

The Pioneers were the first spacecraft to be launched fast enough to escape the gravity of the Solar System. They join the Voyager spacecraft as the only four objects to date that will leave our Solar System and venture into interstellar space.

Instead of a mission using a 16-foot-diameter-high gain antenna, Galileo now had to rely on a tiny low gain antenna the size of a pie pan. Galileo's main antenna had the capability to transmit 134 kb/s from the distant Jupiter, but its low gain antenna could only squeeze out up to 16 bits/s.

Chapter Five

There, the god of the evening star battled the forces of darkness on behalf of humanity. One of the best preserved writings of the Quiche Maya is known as the *Popol Vuh.* This codex documents royal lineages, provides historical accounts, and describes the Mayan creation stories. In one narrative, Hunaphu determines to destroy the lords of darkness. He entertains them with a magic trick. The evil lords watch as Hunaphu kills his twin brother (the Sun god Xbalanque), rips his heart out, then brings him back to life. The delighted onlookers, in effect, say "Me, too!" Hunaphu obliges, but this time he really kills them. He returns home to rise at the first dawn of creation.

Only the prayers of the people and the sacrifices of the high priests could assure Hunahpu's resurrection. Our description here is a simplification of the highly complex Mayan theology. As Marc Levine of the Denver Museum of Nature and Science explains, "Some have argued that Hun Hunahpu (twin brother of Xbalenque, father of Hunahpu) represented Venus in the Mayan Hero Twins creation story. Others affirm that the son, Hunahpu, represents Venus; thus both are interpreted as metaphors for Venus. It is important to remember that the association is affirmed by scholars today but remains open to interpretation."

Later, it was hoped, he would rise as his divine brother, Xbalanque. Unlike deities of Europe and Asia, many gods in the Mayan pantheon have dual identities. The relationships are complex. For more on this, see the excellent resource *Star Gods of the Maya* by Susan Milbrath (University of Texas Press, 1999).

…a technician decided on his own to "improve" the external connections by using shrink tubing. This was not the first time that a technician had "improved" a spacecraft, to the detriment of the mission. The later Soviet Venera landers were equipped with color tiles so that analysts on Earth could calibrate color images transmitted from the surface. The original design of the tiles entailed an exacting process of manufacturing ceramics that would not change color in Venus's blistering heat. But a factory worker realized that the substitution of simple enamel paint would save the project a great deal of time and money. The worker was rewarded for creative thinking. The scientists were rewarded by color images of fried calibration strips that had turned brown instantly, rendering them useless.

Pioneer Venus underlined the fact that the study of planets, in a very real sense, is the study of Earth. In an equally striking link between environmentalism and the space program, astronaut Bill Anders snapped a shot of Earth rising over the lunar horizon during Apollo 8. This emblematic image became a rallying symbol for the environmental movement that began in the late sixties.

Chapter Six

The two-faced nature of the Red Planet is called planetary dichotomy. Our own Moon has a similar dichotomy, with vast plains on its Earth-facing hemisphere and rugged, cratered highlands predominant on the far side. Saturn's two-toned moon Iapetus is yet another example of planetary dichotomy.

A meteor apparently exploded somewhere overhead, peppering the village with some forty stony fragments all over the area. Local Nakhla farmer Mohammad Ali Effendi Hakim claimed that his dog was vaporized by one of the meteor fragments. Since then, the apocryphal report has come under question. Though it was in contemporary newspapers (in both Arabic and English), there was only one eyewitness – the farmer – and no apparent remains of the dog were ever produced. Nevertheless, the Nakhla dog has become an historic celebrity as the first documented death from a meteorite.

Scientists have linked this subtle cycle to Earth's ice ages. During an ice age, the entire Earth cools off, which causes a chain reaction. First, water evaporating from lakes and seas falls on land as snow and ice. Sea levels drop as more and more water is trapped on the surface. The bright ice and snow reflect more sunlight – and heat – back into space, increasing the cooling effect.

Kerzhanovich, who was a Soviet scientist with both the Russian Scientific Research Institute and the Institute of Space Research (IKI)… IKI is Russia's counterpart to the U.S.'s Jet Propulsion Laboratory, NASA's planetary exploration arm.

Chapter Seven

Saturn's belts are harder to see, as they float beneath golden hazes, but it is clear that the ringed giant has a similar number of belts… In fact, all four gas giants have rings, but Saturn's are the most extensive and beautiful.

Chapter Eight

…view of Titan's surface, covered in ice flows and methane fjords, painted shortly after the Voyager Saturn missions. The painting is based on an aerial photo of a fjord in Norway. Unfortunately, I forgot that, unlike water, frozen methane is denser than liquid methane. Any icebergs would be at the bottom of the sea, not floating on top of it! This was diplomatically pointed out by Carl Sagan in a letter to the *Astronomy* magazine editors, where the painting first appeared.

Chapter Nine

Beyond the orbit of Saturn lie two worlds long thought to be identical siblings. Uranus and Neptune are virtually equal in size… In the 1961 *Picture History of Astronomy*, author Patrick Moore is only able to devote three brief paragraphs to Uranus and Neptune. He begins by saying, "Not much is known about the two outer giants, Uranus and Neptune, which are almost perfect twins…They seem to be made on the same pattern as Jupiter and Saturn…" William K. Hartmann's 1978 textbook *Astronomy: The Cosmic Journey* suggests that Neptune "is believed to resemble Uranus in general." Nearly two decades later, National Geographic's *Our Universe* by Roy A. Gallant stated that "Neptune seems to be a twin to Uranus…" Little observational progress was made until later in the Space Age, with the advent of better telescopes and spacecraft reconnaissance. (See Chapter Three.)

Computer simulations indicate that GDS-89 is a vortex turning anticyclonically within the streams of Neptune's surrounding atmosphere. See Polvani et al. in *Science 249*, 1393–1398, 1990.

The storm acts as an obstruction in the wind's flow. As air is deflected upward, methane clouds condense. For a more technical description, see Stratman et al. in *Icarus 151*, 275–285, 2001.

Measuring 2,706 km, it's the seventh-largest moon in the Solar System. From large to small, the largest moons of the Solar System (with their diameters) are: Ganymede (5,265 km), Titan (5,150 km), Callisto (4,810 km), Io (3,636 km), Earth's Moon (3,475 km), Europa (3,138 km) and Triton (2,706 km).

Chapter Ten

…adds Lockheed Martin's Ben Clark. "We're working on a Titan mission now to land in one of the lakes…" See *Titan Mare Explorer TiME): the first exploration of an extraterrestrial sea*, presentation to NASA's Decadal Survey, Space Policy Online.

Chapter Eleven

Geoffrey Landis, researcher at NASA's Glenn Research Center, recently completed a NASA-funded study to see what it would take to set up camp in the clouds of Venus. See *AIP Conference Proceedings Vol. 653*, pp. 1193–1198, from the *Conference on Human Space Exploration*, Space Technology & Applications International Forum, Albuquerque, NM, February 2–6, 2003.

Glossary

Aeronomy The study of the upper region of an atmosphere.

Astronomical unit The distance between Earth and the Sun, about 93 million miles.

Condensation The change from a vapor (a gaseous state) into liquid.

Coriolis effect Apparent deflection of the motion of atmosphere as the planet rotates under it.

Evaporation Vaporization of liquid into gas.

Exosphere Uppermost layer of an atmosphere.

Hadley cell A loop of atmospheric circulation where heated air rises over the equator, moves poleward, and sinks again.

Hydrostatic equilibrium The force of atmospheric pressure acting against the force of gravity to keep an atmosphere stable.

Mesosphere Portion of Earth's atmosphere 30–50 miles above the surface, where temperature decreases with altitude.

Occultation The observation of an object as it passes in front of another. Astronomers use occultations to measure structure and density of atmospheres.

Photodissociation The process in which sunlight or other radiation splits molecules of gas.

Rossby waves Undulations in the upper atmosphere that cause condensation and mixing of the air.

Sidereal day The time it takes for a planet to make one rotation around its spin axis compared to the background stars.

Solar day The time it takes for the Sun to return to the same apparent spot in the sky.

Stratosphere Calm region of air above the troposphere, extending up to about 35 miles.

Sublimate The process by which material in a frozen state turns directly into gas without going through a liquid phase.

Thermosphere The most extensive portion of Earth's atmosphere, between the mesosphere and exosphere.

Troposphere Lowest and most dense part of the atmosphere, region where nearly all of the weather phenomena takes place.

UAV Unmanned aerial vehicle.

Index

Printed in the United States of America